Circular Economy and Sustainable Energy Materials

Achieving carbon neutrality is crucial for creating a sustainable and resilient future worldwide. The circular economy framework focuses on reducing waste, optimizing resource use, and promoting regenerative practices, thus curbing carbon footprints. Concurrently, sustainable energy techniques, such as harnessing renewable sources, enhancing energy storage, and boosting energy efficiency, contribute to the reduction of greenhouse gas emissions. With a unique emphasis on net-zero emission approaches, this book delves into circular economy principles and sustainable energy materials, offering a comprehensive perspective on climate change challenges.

- Covers fundamental circular economy principles, from carbon capture to advancements in biomass and hydrogen energy
- Explores recycling techniques for essential energy materials, including batteries, solar cells, and fuel cells
- Provides detailed coverage of technologies, processes, and challenges related to achieving sustainability in the energy sector
- Bridges theory and practical applications

Offering a roadmap toward a carbon neutral and net-zero emission future, this book serves as a valuable resource for advanced students, researchers, engineers, and policymakers worldwide seeking solutions to climate change and sustainable development.

Circular Economy and Sustainable Energy Materials

A Net-Zero Emissions Approach

Edited by
Ngoc Thanh Thuy Tran
Wei-Sheng Chen
Shou-Heng Liu
Jow-Lay Huang

CRC Press
Taylor & Francis Group
Boca Raton London New York

CRC Press is an imprint of the
Taylor & Francis Group, an **informa** business

Cover image: Miha Creative/Shutterstock.com

First edition published 2025
by CRC Press
2385 NW Executive Center Drive, Suite 320, Boca Raton FL 33431

and by CRC Press
4 Park Square, Milton Park, Abingdon, Oxon, OX14 4RN

CRC Press is an imprint of Taylor & Francis Group, LLC

ISBN: 978-1-032-80356-2 (hbk)
ISBN: 978-1-032-80357-9 (pbk)
ISBN: 978-1-003-46300-9 (ebk)

DOI: 10.1201/9781003463009

Typeset in Times
by SPi Technologies India Pvt Ltd (Straive)

Preface

Achieving carbon neutrality is crucial for creating a sustainable and resilient future worldwide. As the impacts of climate change intensify, transitioning to a state of carbon neutrality becomes an imperative to mitigate the adverse effects on the environment. Circular economy principles and sustainable energy techniques are vital in this transformative journey. The circular economy framework focuses on reducing waste, optimizing resource use, and promoting regenerative practices, thus curbing carbon footprints. Concurrently, sustainable energy techniques, such as harnessing renewable sources, enhancing energy storage, and boosting energy efficiency, contribute to the reduction of greenhouse gas emissions. This book delves into circular economy principles and sustainable energy materials, offering a comprehensive perspective on climate change challenges.

With a unique emphasis on net-zero emission approaches, the book covers fundamental circular economy principles, from carbon capture to advancements in biomass and hydrogen energy. Recycling techniques for essential energy materials, including batteries, solar cells, and fuel cells, are also explored. It provides a detailed exploration of technologies, processes, and challenges related to achieving sustainability in the energy sector. Each chapter bridges theory and practical applications, providing insights to guide us towards a future of carbon neutrality.

This book is a testament to the collaborative efforts of contributors who share a vision of a world where circular economy is not merely a concept but a guiding principle woven into the fabric of our sustainable environment. As we stand at the intersection of environmental stewardship and technological innovation, this book aims to provide a roadmap towards a carbon neutrality and net-zero emission future. It provides a valuable resource for students, researchers, engineers, and policymakers worldwide seeking solutions to climate change and sustainable development.

Acknowledgements

This work was financially supported by the Hierarchical Green-Energy Materials (Hi-GEM) Research Center, from The Featured Areas Research Center Program within the framework of the Higher Education Sprout Project by the Ministry of Education (MOE) in Taiwan.

Contents

Editors

Ngoc Thanh Thuy Tran obtained her PhD in physics from the National Cheng Kung University (NCKU), Taiwan, in 2017. Following her doctoral studies, she embarked on a career as a postdoctoral researcher and later assumed the role of an Assistant Researcher at the Hierarchical Green-Energy Materials (Hi-GEM) Research Center at NCKU. Dr. Tran's scientific pursuits primarily revolve around investigating the fundamental (electronic, magnetic, and thermodynamic) properties of 2D materials and rechargeable battery materials, employing first-principles calculations.

Wei-Sheng Chen serves as an Associate Professor in the Department of Resources Engineering at National Cheng Kung University, Taiwan. His research interests encompass circular economy, metal recycling, waste management, and zero-emission practices.

Shou-Heng Liu holds the position of Professor in Department of Environmental Engineering at National Cheng-Kung University, Taiwan. His research focuses on clean energy technology, environmental electrochemistry, environmental nanocatalysis, and sustainable resources engineering.

Jow-Lay Huang is a Chair Professor in the Department of Materials Science and Engineering at National Cheng Kung University, Taiwan. Additionally, he serves as the Director of the Hierarchical Green-Energy Materials (Hi-GEM) Research Center at NCKU. Dr. Huang earned his PhD in material science and engineering in 1983 from the University of Utah, Salt Lake City, Utah, USA. His diverse research portfolio encompasses the fabrication, development, and application of ceramic nanocomposites, piezo-phototronic thin films for photodetector devices, piezoelectric thin films for high-frequency devices, and metal oxide/graphene and SiCx nanocomposites as anode materials for lithium-ion batteries. He is also actively involved in exploring the potential of 2D nanocrystal materials for photoelectrochemical applications.

Contributors

Han-Yi Chen
Department of Materials Science and Engineering
National Tsing Hua University
Hsinchu City, Taiwan

Wei-Sheng Chen
Department of Resources Engineering
National Cheng Kung University
Tainan City, Taiwan

Wei-Ting Chen
Institute of Environmental Engineering
National Sun Yat-sen University
Kaohsiung City, Taiwan

Wen-Hui Cheng
Department of Materials Science and Engineering
National Cheng-Kung University
Tainan City, Taiwan

Kuan-Zong Fung
Hierarchical Green-Energy Materials (Hi-GEM) Research Center
National Cheng Kung University
Tainan City, Taiwan

Hsing-Jung Ho
Department of Environmental Studies for Advanced Society
Graduate School of Environmental Studies
Tohoku University
Sendai City, Japan

Jow-Lay Huang
Hierarchical Green-Energy Materials (Hi-GEM) Research Center
and
Department of Materials Science and Engineering
and
Center for Micro/Nano Science and Technology
National Cheng Kung University
Tainan City, Taiwan

Atsushi Iizuka
Department of Environmental Studies for Advanced Society
Graduate School of Environmental Studies
Tohoku University
Sendai City, Japan

Chih-Ming Kao
Institute of Environmental Engineering
National Sun Yat-sen University
Kaohsiung City, Taiwan

Yi-Hsuan Lai
Department of Materials Science and Engineering
National Cheng-Kung University
Tainan City, Taiwan

Nguyen Viet Linh Le
Department of Environmental Engineering
National Cheng Kung University
Tainan City, Taiwan

Cheng-Han Lee
Department of Resources Engineering
National Cheng Kung University
Tainan City, Taiwan

Jin-Min Li
Institute of Environmental Engineering
National Sun Yat-sen University
Kaohsiung, Taiwan

Chia-Yu Lin
Department of Chemical Engineering
National Cheng Kung University
Tainan City, Taiwan

Shou-Heng Liu
Department of Environmental Engineering
National Cheng Kung University
Tainan City, Taiwan

Jiun-Hau Ou
Institute of Environmental Engineering
National Sun Yat-sen University
Kaohsiung City, Taiwan

Fitri Nur Indah Sari
Department of Chemical Engineering
National Cheng Kung University
Tainan City, Taiwan

Ngoc Thanh Thuy Tran
Hierarchical Green-Energy Materials (Hi-GEM) Research Center
National Cheng Kung University
Tainan City, Taiwan

Shu-Yi Tsai
Hierarchical Green-Energy Materials (Hi-GEM) Research Center
National Cheng Kung University
Tainan City, Taiwan

Meng-Wei Zheng
Department of Environmental Engineering
National Cheng Kung University
Tainan City, Taiwan

1 Introduction

Ngoc Thanh Thuy Tran, Shou-Heng Liu, Wei-Sheng Chen, Chia-Yu Lin, and Wen-Hui Cheng
National Cheng Kung University, Tainan City, Taiwan

Han-Yi Chen
National Tsing Hua University, Hsinchu City, Taiwan

Chih-Ming Kao
Institute of Environmental Engineering, National Sun Yat-sen University, Kaohsiung City, Taiwan

Shu-Yi Tsai
National Cheng Kung University, Tainan City, Taiwan

Hsing-Jung Ho
Tohoku University, Sendai City, Japan

Jow-Lay Huang
National Cheng Kung University, Tainan City, Taiwan

Welcome to the thought-provoking journey through the *Circular Economy and Sustainable Energy Materials: Net-zero Emission Approach.* In this book, each chapter contributes to a comprehensive exploration of sustainable energy materials within the framework of a circular economy. The journey begins with an insightful 'Introduction' in Chapter 1, setting the stage for the subsequent chapters. Chapter 2 delves into the principles of 'Circular Economy,' establishing a foundation for sustainable resource management. Chapters 3 and 4 focus on 'CO_2 Capture' and 'Mineral Carbonation' as crucial strategies for carbon sequestration and utilization. Advancing from these foundational principles, Chapters 5 and 6 explore innovative technologies, such as 'Photoelectrochemical Systems' and 'Nanophotonic Architectures,' for upcycling carbon-containing waste and reducing CO_2 emissions. Chapter 7 introduces the application of 'Water Chestnut Husks-derived Activated Carbons' as electrodes, highlighting bio-based materials for energy storage. Moving forward, Chapters 8–11 cover various aspects of sustainable energy, including 'Biomass Energy,' and 'Hydrogen Energy.' Chapters 12–14 focus on "Recycling Batteries," "Solar Cell Materials,", and "Fuel Cell Materials," emphasizing the importance of closing material loops. The journey concludes with Chapter 15, providing insightful Concluding

DOI: 10.1201/9781003463009-1

Remarks that tie together the diverse threads of circular economy and sustainable energy, offering a holistic perspective on achieving net-zero emissions. This book serves as a comprehensive exploration of innovative strategies, technologies, and practices aimed at reshaping our approach to resource management, waste reduction, and sustainable energy utilization.

The circular economy (CE) is defined as the sustainable management of resources by minimizing waste and promoting the continuous reuse and recycling of products and materials. This innovative economic model not only aims to reduce pressure on limited raw materials, but also holds great promise in the pursuit of net-zero emissions. With the introduction of these goals, countries worldwide have embarked on the "2030 Agenda" to achieve these objectives by 2030. At the same time, CE has also become the target of joint efforts of all countries due to the temperature targets set in the 2015 Paris Agreement. A series of strategies underpin this model, including reimagining product design, optimizing resource efficiency and fostering a culture of recycling and material reuse. These strategies support the transition to a more environmentally friendly and resource-saving economic system, ensuring a harmonious balance between economic growth and ecological protection. Chapter 2 explains the interrelationship between the CE and net zero emissions and examines the opportunities presented by the CE using different strategies.

Carbon dioxide (CO_2) is a greenhouse gas that plays a significant role in climate change. When released into the atmosphere, CO_2 acts as a heat-trapping blanket, preventing some of the sun's heat from escaping back into space. This phenomenon, known as the greenhouse effect, is essential for maintaining a habitable temperature on Earth. However, human activities, primarily the burning of fossil fuels such as coal, oil, and natural gas, have dramatically increased the concentration of CO_2 in the atmosphere. The increased levels of CO_2 and other greenhouse gases in the atmosphere have led to an intensification of the greenhouse effect, resulting in global warming and various impacts on the climate system. To reduce the hazards of CO_2, carbon capture has been implemented in the last few years. Carbon capture is a process aimed at reducing CO_2 emissions from various sources, particularly large-scale industrial facilities such as power plants and factories. It involves capturing CO_2 emissions before they are released into the atmosphere, transporting the captured CO_2 to a storage location, and securely storing it for an extended period of time, as discussed in Chapter 3.

Mineral carbonation is one of the carbon capture, utilization, and storage technologies that capture CO_2 and convert it into inorganic carbonated minerals as a product for further utilization [1]. CO_2 gas will react with Ca/Mg from alkaline materials to form $CaCO_3/MgCO_3$. In addition to natural silicate rocks, industrial waste can also be raw material to provide contribution on CO_2 emissions reduction. Industrial waste and byproducts are more active and cheaper than natural rocks, indicating industrial waste and byproducts are promising candidate for mineral carbonation [2–5]. The concept of mineral carbonation of alkaline waste will be introduced. Various processes can be developed according to different purposes, target materials, and target industries. In Chapter 4, these considerations will be described [6, 7]. The role of mineral carbonation using alkaline waste and byproducts in CO_2 emissions reduction will be explained by a case evaluation in Taiwan [8]. Based on the concept of mineral carbonation of alkaline waste, a great contribution for industries to move towards carbon neutrality and a circular economy can be achieved.

Around 70% of global municipal solid waste, mainly consisting of food and biomass (~46%), paper and cardboard (17%), and plastics (~17%), accumulate in landfills or are even discharged into our environment every year [9], which poses threats to the public health, environment, and aquatic ecosystem. Plastic waste is of particular concern due to its non-biodegradability and associated persistent pollution. It is anticipated that global greenhouse gas emissions can be reduced by up to 20% if this solid waste can be prevented, recovered, and recycled [10]. Instead of being sent to landfills or acting as pollution sources, most of these solid wastes are of high energy density and can be attractive feedstocks for the synthesis of commodity chemicals and clean fuels. Photo-(electro-)catalytic reforming utilizes waste and water as the feedstocks for the generation of valuable chemicals (e.g., formic acid, 2,5-furandicarboxylic acid) and fuels (e.g., hydrogen), and has been considered as the promising approach to address challenges posed by increasing global energy demand and environmental pollution. In Chapter 5, we will examine the recent progress and remaining challenges in the development of photoelectrocatalytic reforming systems for upcycling biomass and plastic waste into value-added chemicals.

The goal of Chapter 6 is to address innovation of tuning light-matter interaction for energy application. Fundamentals of photo-driven CO_2 reduction will be addressed. Also, design principles, including material selection and optical theory, are included. Lastly, current experimental demonstrations and approaches are compared and discussed. The enhancement of the overall efficiency and product selectivity will make the platform more promising for unassisted solar fuel generation, especially from CO_2. The development of simulation and analysis will pave the route for mechanism development of the photoelectrochemical reaction.

In recent years, there has been growing interest in biomass-derived materials for energy storage and power generation. Water chestnut shells, which are a sustainable resource, are now favored as unique carbon-based materials. Converting this abundant agricultural waste into cost-effective carbon electrode materials can help establish a sustainable circular economy. Chapter 7 explores the versatile applications of these materials, emphasizing their role as low-cost sustainable electrodes and their potential in supercapacitors. Synthesis methods, sodium- and potassium-ion batteries, microbial fuel cells, and plant microbial fuel cells were also discussed. These materials offer ecofriendly power generation, contributing to cleaner and more efficient energy solutions while addressing environmental challenges.

The demand for energy has also increased strongly in the last few decades. Extensive utilization of traditional fuel sources such as fossil fuels have caused significant environmental impact (i.e., global warming effect). Among alternative fuels, biomass energy is considered an enormously potential fuel source [11–13]. Biomass, the main source of raw materials for biofuel production, is available almost everywhere and even underutilized. All sources of microorganisms, plants or animals are biomass and have potential for biofuel production. Agricultural residues and waste from industry, farms and households are the largest sources of biomass. The utilization of these by-products and waste sources can not only unravel the problem of energy but also solve the environmental problems, resulting in the complete implementation of the economic circle. In Chapter 8, the process optimization of converting biomass into energy suitable for many different biomass sources is discussed in order to achieve the requirements for applications on an industrial scale.

Renewable energy in recent years has become one of the scorching topics when the problems of air pollution from traditional fuel consumption could not be significantly improved. Among them, hydrogen, a fuel source that does not contain carbon and can be used directly in internal combustion engines without requiring engine modification, is one of the extremely promising fuel sources [14–17]. Because fossil fuel is not a renewable fuel source, gray hydrogen is also not considered a renewable fuel source. Unfortunately, more than 95% of all hydrogen produced worldwide depends on fossil fuel, and gray and blue hydrogen account for half of that. In the opposite direction, green hydrogen generated from renewable energy sources such as solar energy, wind energy, or biomass resources is considered extremely eco-friendly. As such, they not only have the potential to achieve the goal of carbon neutrality but also solve other serious energy problems. However, the approach is still emerging and the development of facilities as well as technology have not kept pace with energy demand. Within the scope of Chapter 9, the production and application aspects of green hydrogen are discussed since it is the only hydrogen source that can help create a sustainable and circular economy and achieve net-zero emission goal.

In today's society, the rapid development of technology is constantly changing our way of life and social structure. The demand for science and technology is increasing, and understanding these fields is becoming more and more important. In this preface, we would like to introduce a technology for sustainable energy use: reversible solid oxide cell (rSOC) as provided in Chapter 10. rSOC operates alternately under SOFC and SOEC modes to realize the cyclicality of sustainable energy use of "hydrogen energy-power generation" [18–20]. rSOC is a technology that can convert fuel gases, such as hydrogen, into electricity in the SOFC mode, and convert electricity back into hydrogen or other storable fuels in the SOEC mode. This cyclic operation of rSOC enables efficient energy conversion and storage, establishing a sustainable loop between energy supply and utilization. By alternating between the SOFC and SOEC modes, rSOC can flexibly adjust the conversion between hydrogen and electricity based on energy demand and supply, enabling the utilization and storage of sustainable energy. This technology plays a crucial role in achieving clean energy systems and facilitating the energy transition.

Chapter 11 highlights that in the biological remediation methods of chlorinated hydrocarbons (CHCs), the reductive dechlorination effect of CHCs is enhanced by immobilizing Dehalococcoides mccartyi BAV1 and Clostridium butyricum in a silicone carrier [21, 22]. Clostridium butyricum, which can produce high hydrogen efficiency through the appropriate addition of carbon sources and has endogenous spores that can withstand extreme environments, shows potential for flexible application. It has a lower carbon footprint by generating hydrogen biologically, compared to the past use of in-situ hydrogen injection or increasing H_2 in water by reacting nZVI with water. The produced hydrogen concentration can reach the required hydrogen threshold for reductive groundwater dechlorination (2.2 ± 0.9 nM). The pilot test used an innovative immobilized Clostridium butyricum (ICB) column scheme. This scheme remediated *cis*-1,2-dichloroethene polluted groundwater through anaerobic reductive dechlorination processes. The results showed that ICB combined with slow-release carbon sources can effectively stimulate the growth of reductive dechlorination-related bacterial communities. The application of ICB can also effectively increase the number of Dehalococcoides and populations with reductive dehalogenase genes, enhancing the dechlorination effect of cis-1,2-dichloroethene and vinyl chloride, making the remediation process greener and more environmentally friendly, and more in line with the goals of green and sustainable remediation. Introducing Clostridium butyricum into the remediation site can be a potential choice for in-situ remediation.

Batteries are electrochemical devices that store and release electrical energy through chemical reactions. They are widely used in various applications, including portable electronics, electric vehicles, renewable energy storage, and backup power systems. Batteries consist of one or more electrochemical cells, each containing two electrodes (anode and cathode) separated by an electrolyte [23–25]. When a battery is in use, a chemical reaction occurs at the electrodes, causing the flow of electrons through an external circuit, generating an electric current. This process converts chemical energy into electrical energy. Different types of batteries employ various chemistries, such as zinc-carbon, mercury, lithium-ion, lead-acid batteries, and others. Each has its unique characteristics, energy density, and performance. When batteries have reached the end of their useful life and can no longer provide adequate electrical power, they will turn into waste batteries. Battery disposal is a significant environmental concern due to the hazardous materials they contain. Many batteries contain toxic substances like lead, mercury, cadmium, and other heavy metals, posing risks to human health and the environment if improperly handled. Therefore, recycling materials from waste batteries helps reduce the environmental impact of battery waste and conserves valuable resources, as investigated in Chapter 12.

Solar cells, or photovoltaic cells, are devices that convert sunlight directly into electricity. They comprise various materials, including silicon, metals (copper, silver, tin, indium, gallium, selenium, cesium, or lead), and glass [26–30]. The growing adoption of solar energy as a renewable power source has led to increased production of different types of solar panels (first generation to third generation) and, consequently, an increased need for proper recycling techniques for solar cell materials. Recycling is essential to address solar panel waste's environmental impact and recover valuable resources. As shown in Chapter 13, by recycling solar cell materials, we can minimize the depletion of natural resources, reduce energy consumption associated with mining and manufacturing, and prevent the accumulation of electronic waste. Furthermore, recycling techniques enable the recovery of valuable metals and other materials, promoting resource efficiency and contributing to a more sustainable and circular economy.

Fuel cells are innovative and environmentally-friendly devices that convert chemical energy directly into electrical energy. They have gained considerable attention as a promising alternative to traditional fossil fuel-based power sources, thanks to their high efficiency and low greenhouse gas emissions [31–38]. These advanced energy conversion systems employ catalysts, often containing precious metals and rare earth elements, to facilitate the electrochemical reactions that produce electricity. As the world embraces sustainable practices, the importance of recycling materials from fuel cells becomes increasingly evident (They would also face the degradation problem). Recycling these valuable components not only reduces the environmental impact of their production and disposal but also ensures the preservation of critical resources. In this context, exploring the recycling methods for fuel cells promotes a cleaner and greener energy landscape and contributes to building a more sustainable and resource-efficient future, which is addressed in Chapter 14.

REFERENCES

[1] Ho H-J, Iizuka A, Shibata E 2019. Carbon capture and utilization technology without carbon dioxide purification and pressurization: A review on its necessity and available technologies. *Ind. Eng. Chem. Res.* 58(21):8941–54. doi:10.1021/acs.iecr.9b01213

[2] Ho H-J, Iizuka A, Shibata E, Tomita H, Takano K, Endo T 2021. Utilization of CO_2 in direct aqueous carbonation of concrete fines generated from aggregate recycling: Influences of the solid–liquid ratio and CO_2 concentration. *J. Clean. Prod.* 312(January):127832. doi:10.1016/j.jclepro.2021.127832

[3] Ho H-J, Iizuka A, Shibata E, Tomita H, Takano K, Endo T 2020. CO_2 utilization via direct aqueous carbonation of synthesized concrete fines under atmospheric pressure. *ACS Omega* 5(26):15877–90. doi:10.1021/acsomega.0c00985

[4] Ho H-J, Iizuka A, Kubo H 2022. Direct aqueous carbonation of dephosphorization slag under mild conditions for CO_2 sequestration and utilization: Exploration of new dephosphorization slag utilization. *Environ. Technol. Innov.* 28:102905. doi:10.1016/j.eti.2022.102905

[5] Ho H-J, Iizuka A, Shibata E 2021. Utilization of low-calcium fly ash via direct aqueous carbonation with a low-energy input: Determination of carbonation reaction and evaluation of the potential for CO_2 sequestration and utilization. *J. Environ. Manag.* 288(February):112411. doi:10.1016/j.jenvman.2021.112411

[6] Izumi Y, Iizuka A, Ho H-J 2021. Calculation of greenhouse gas emissions for a carbon recycling system using mineral carbon capture and utilization technology in the cement industry. *J. Clean. Prod.* 312(December 2020):127618. doi:10.1016/j.jclepro.2021.127618

[7] Ho H, Iizuka A, Shibata E, Ojumu T 2022. Circular indirect carbonation of coal fly ash for carbon dioxide capture and utilization. *J. Environ. Chem. Eng.* 10(5):108269. doi:10.1016/j.jece.2022.108269

[8] Ho H-J, Iizuka A, Lee C-H, Chen W-S 2023. Mineral carbonation using alkaline waste and byproducts to reduce CO_2 emissions in Taiwan. *Environ. Chem. Lett.* 21(2):865–84. doi:10.1007/s10311-022-01518-6

[9] Uekert T, Pichler CM, Schubert T, Reisner E 2021. Solar-driven reforming of solid waste for a sustainable future. *Nat. Sustain.* 4(5):383–91.

[10] ISWA, UNEP 2015.Global waste management outlook (United Nations Environment Programme), *International Solid Waste Association.*

[11] Prusak R., Skuza Z., Kurtyka M., Rembiesa Z. 2018. Potential of biomass-to-fuel conversion technologies for power and means of transport. *J. KONES*, 25(2): 287–294.

[12] Bridgwater T 2006. Biomass for energy. *J. Sci. Food Agric.* 86(12):1755–68. doi:10.1002/JSFA.2605

[13] Bich T, Nguyen N, Viet N, Le L 2023. Biomass resources and thermal conversion biomass to biofuel for cleaner energy: A review. *J. Emerg. Sci. Eng.* 1(1):6–13. doi:10.61435/JESE.2023.2

[14] Atilhan S, Park S, El-Halwagi MM, Atilhan M, Moore M, Nielsen RB 2021. Green hydrogen as an alternative fuel for the shipping industry. *Curr. Opin. Chem. Eng.* 31:100668. doi:10.1016/J.COCHE.2020.100668

[15] Olateju B, Kumar A 2011. Hydrogen production from wind energy in Western Canada for upgrading bitumen from oil sands. doi:10.1016/j.energy.2011.09.045

[16] Olateju B, Kumar A, Secanell M. 2016. A techno-economic assessment of large scale wind-hydrogen production with energy storage in Western Canada. doi:10.1016/j.ijhydene.2016.03.177
[17] Hydrogen on the path to net-zero emissions. Accessed: Jun. 07, 2023. [Online]. Available: https://about.jstor.org/terms
[18] Li HY, Xu YX, Lv N, Zhang QYT, Zhang XL, Wei ZJ, Wang YD, Tang HL, Pan HF 2023. Ti-doped SnO_2 supports IrO_2 electrocatalysts for the oxygen evolution reaction (OER) in PEM water electrolysis. *ACS Sustain. Chem. Eng.* 11:1121–32.
[19] Rheinlander PJ, Durst J 2021. Transformation of the OER-active IrOx species under transient operation conditions in PEM water electrolysis. *J. Electrochem. Soc.* 68:024511.
[20] Wang Y, Pang YH, Xu H, Martinez A, Chen KS 2022. PEM fuel cell and electrolysis cell technologies and hydrogen infrastructure development - A review. *Energy Environ. Sci.* 15:2288–28.
[21] Dutta N, Usman M, Ashraf MA, Luo G, Zhang S 2022. A critical review of recent advances in the bio-remediation of chlorinated substances by microbial dechlorinators. *Chem. Eng. J. Adv.* 12:100359.
[22] Xiao Z, Jiang W, Chen D, Xu Y 2020. Bioremediation of typical chlorinated hydrocarbons by microbial reductive dechlorination and its key players: A review. *Ecotoxicol. Environ. Saf.* 202:110925.
[23] Tran NTT, Hsu WD, Huang JL, Lin MF (Eds.) (2022). *Lithium-Related Batteries: Advances and Challenges*. CRC Press.
[24] Tran NTT, Lin CA, Lin SK 2023. Insights into the structural and thermodynamic instability of Ni-rich NMC cathode. *ACS Sustain. Chem. Eng.* 11(18):6978–87.
[25] Nguyen TDH, Lin SY, Chung HC, Tran NTT, Lin MF 2021. *First-Principles Calculations for Cathode, Electrolyte and Anode Battery Materials*. IOP Publishing.
[26] Al-Ezzi AS, Ansari MNM 2022. Photovoltaic solar cells: A review. *Appl. Syst. Innov.* 5:67.
[27] Breitenstein O 2013. The physics of industrial crystalline silicon. *Solar Cells* 89:1–75.
[28] Tian H, Sun L 2013. *Organic Photovoltaics and Dye-Sensitized Solar Cells*. Elsevier B. V. 567–605
[29] Pastuszak J, Węgierek P 2022. Photovoltaic cell generations and current research directions for their development. *Materials* 15:554.
[30] Tu LH, Tran NTT, Lin SK, Lai CH 2023. Efficiency boost of $(Ag_{0.5}, Cu_{0.5})(In_{1-x}, Ga_x)Se_2$ thin film solar cells by using a sequential process: effects of ag-front grading and surface phase engineering. *Adv. Energy Mater.* 13(29):2301227.
[31] Pan ZF, An L, Wen CY 2019. Recent advances in fuel cells based propulsion systems for unmanned aerial vehicles. *Appl. Energy* 240:473–85.
[32] Edwards PP, Kuznetsov VL, David WIF, Brandon NP 2018. Hydrogen and fuel cells: Towards a sustainable energy future. *Energy Policy* 36:4356–62.
[33] Neef H-J 2009. International overview of hydrogen and fuel cell research. *Energy* 34:327–33.
[34] Carrette L, Friedrich K A, Stimming U 2000. Fuel cells: Principles, types, fuels, and applications. *ChemPhysChem* 1:162–93.
[35] Carrette L, Friedrich K A, Stimming U 2001. Fuel cells - Fundamentals and applications. *Fuel Cells* 1:5–39.
[36] Garche J, Jurissen L 2015. Applications of fuel cell technology: Status and perspectives. *Electrochem. Soc. Interface* 24:39–43.
[37] Luo Y, Wu Y, Li B, Mo T, Li Y, Feng S-P, et al. 2021. Development and application of fuel cells in the automobile industry. *J. Energy Storage* 42:103124.
[38] Perčić M, Vladimir N, Jovanović I, Koričan M 2022. Application of fuel cells with zero-carbon fuels in short-sea shipping. *Appl. Energy* 309:11846.

2 Principles and Strategies of Circular Economy

Meng-Wei Zheng and Shou-Heng Liu
National Cheng Kung University, Tainan City, Taiwan

The concept of circular economy (CE) revolves around the sustainable management of resources by minimizing waste and promoting the continuous reuse and recycling of products and materials. This innovative economic model not only aims to reduce pressure on limited raw materials, but also holds great promise in the pursuit of net-zero emissions. Policies related to agriculture and industry play a key function in reducing carbon emissions and promoting sustainable development by resource management, waste reduction, and integration of renewable energy. A series of strategies underpin this model, including reimagining product design, optimizing resource efficiency and fostering a culture of recycling and material reuse. These strategies support the transition to a more environmentally friendly and resource-saving economic system, ensuring a harmonious balance between economic growth and ecological protection.

2.1 INTRODUCTION

The growth of the global population has led to high resource consumption, further exacerbating various challenges, such as the widening gap between rich and poor communities. A substantial amount of literature on various aspects of the CE had been published, including its conceptualization, sectoral applications, business models, ecosystems, financing, policies, behaviors, and more in recent years. The CE is conceptualized as an industrial system that intentionally seeks to restore or regenerate value. It is a desire to restore value, transition to the use of renewable energy, avoid the use of hazardous chemicals in recycling processes, and aim to eliminate waste via superior material design in products and systems (1). With the introduction of these goals, countries worldwide have embarked on the "2030 Agenda" to achieve these objectives by 2030. At the same time, CE has also become the target of joint efforts of all countries due to the temperature targets set in the 2015 Paris Agreement (2, 3). This chapter explains the interrelationship between the CE and net zero emissions and examines the opportunities presented by the CE using different strategies.

2.2 CONCEPT OF CE

CE represents paradigm shift in the use of natural resources. At its core, CE emphasizes the closed-loop flow of materials, reducing and minimizing inputs (raw materials, water, and energy) and unnecessary outputs (waste and emissions) across multiple

DOI: 10.1201/9781003463009-2

stages (4). To achieve this target, a comprehensive approach that links upstream resource efficiency with downstream waste and emission issues is required. Moriguchi emphasizes that the value of products, materials, and services must be maintained for as long as possible, as covered by the principles of CE (5). This implies a shift towards a circular economic model where "reduce, reuse, and recycle" are key to maximizing circularity.

In the previous literature, various approaches are found based on CER strategies. The main difference between these methods is the number of R, included in this model (Figure 2.1) (6) as follows:

1. Reduce Input Consumption (RIC): RIC focuses on processes such as redesigning original products, reducing the occurrence of redundant products, creating multi-functional products, minimizing manufacturing resources, and promoting more intensive use through product sharing.
2. Reuse: The goal of this R is to emphasize the reuse of products. Even after discarding, consumers can reprocess them to restore their original functions and directly extend the service life of the product.
3. Recover: This R refers to ways to extend the life of products and parts. For example: repair or maintenance of defective products, use of related components in similar products, use of obsolete products or components in other types of products, and repair or reuse of old products.
4. Recycle: This R obtains high-quality materials to replace the use of natural resources and realize material recycling applications.
5. Reduce Waste and Emissions (RWE): RWE is an outcome of other R's and should prioritize resource recovery. In the final stages of this process, CE requires the formulation of waste and emissions management strategies, which should seek to enhance resource recycling while minimizing environmental consequences.

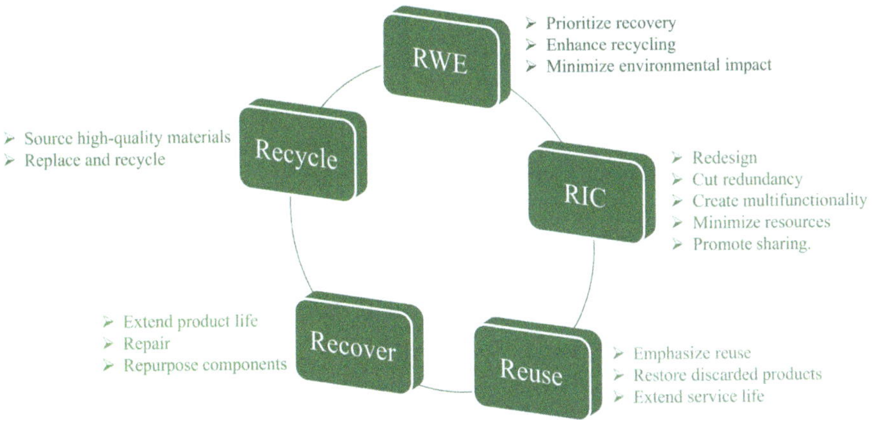

FIGURE 2.1 Circular economy with the "5Rs".

These principles collectively form the foundation of CE strategies, emphasizing sustainable development, resource recycling, and environmental responsibility. This section analyzes the impact of CE strategies on various applications, including energy consumption, natural resources, air and water quality, and waste management, as well as land use and coverage.

2.2.1 Energy Consumption

Energy consumption is one of the major global challenges. Many countries and multinational companies have solved the problem of energy consumption from the perspective of CE and achieved sustainable development. Padmanaban et al. recover energy from recycled plastics as alternative fuels via the goal of CE. A portion of the organic waste undergoes anaerobic fermentation to produce biogas for the power plant. In addition, dry waste can be incinerated in power plants to replace grid power supply. Non-recyclable, high calorific value fractions can be used as waste-derived fuel in cement plants, replacing fossil fuels and reducing energy consumption. It can reduce fuel consumption by 7.8% and increase thermal efficiency by 4.7% (7). Laskurain-Iturbe et al. conducted a case study on the CE and found that many technologies can effectively reduce the energy consumption of the CE, including big data analysis, internet of things, network security, computer vision, virtual reality, artificial intelligence, and robots, etc. Combining these strategies can significantly reduce energy consumption in the transportation, manufacturing, and use stages (8).

2.2.2 Natural Resources

The depletion of natural resources is a result of factors such as population growth and economic development. Therefore, the concept of developing a CE has been introduced to simultaneously safeguard natural resources and promote economic growth (9). CE measures have a positive impact in industries heavily reliant on natural resources, leading to a reduction or even optimization of resource consumption (10). Pavalova et al. presented a conceptual model for the CE to speculate on the optimal utilization of natural resources. This model involves the partial replacement of primary materials with secondary materials, significantly reducing the consumption of natural resources (11). Abad-Segura et al. conducted a dual-system and literature-based investigation of research published in 2021 and found that once the integration of CE and bioeconomy is implemented, the efficiency of natural resource utilization can be improved through a systematic approach. On the other hand, policymakers advocate alternative approaches to the CE (acquisition, manufacturing, disposal) to mitigate the extensive consumption of natural resources. Industries dependent on natural resources can adopt CE strategies to reduce resource inputs and improve the efficiency of natural resource utilization through systemic methods (12).

2.2.3 Air and Water Quality

Air pollution and deteriorating water quality have become some of the most widespread global issues, driven by economic growth and population expansion. Concerns about the quality of air and water resources lead to discussions centered around the development of the CE. Su and Urban conducted scenario analysis for a town in eastern China with a population of 140,000, and their simulation results for 2040 indicate that it is feasible to reduce particulate matter (PM2.5) emissions. However, the total emissions for 2040 show no significant change compared to 2019, suggesting that CE strategies do not actually improve air quality but rather prevent further deterioration. In terms of transportation, integrating the CE with electrified vehicles can reduce PM2.5 emissions by 22% and significantly improve air quality (13). Sgroi et al. proposed that new policies employing the concept of the CE could lead to "paradigm shift" towards a more sustainable wastewater management model. Implementing CE strategies that focus specifically on source separation and water reuse can have a significant impact. Reclaimed water ensures safety condition and conforms to relevant water quality standards, ultimately enhancing resource recovery and improving the quality of water systems (14, 15). Additionally, the practice of the CE can adjust and monitor the levels of wastewater pollutants, thus improving the quality of the water supply through optimized wastewater management. In summary, effective CE solutions can significantly enhance the quality of the air and water environments.

2.2.4 Wastes

Solid waste containing potential toxic elements is generated extensively worldwide, including urban sludge, municipal solid waste incineration fly ash, tailings, metallurgical and chemical solid waste, as well as electronic waste (16, 17). Product design and business models based on CE strategies emphasize multi-functional products and extend the service life of products and components. Reduce waste in the public and private sectors by maximizing product utility by smart manufacturing (18). Al-Wahaibi et al. conducted techno-economic study on the potential of producing biogas from food waste. The results showed that food solid waste samples produced a cumulative gas yield of 1.5 L per gram of dry matter, equivalent to $0.39 per cubic meter, with a discounted period of 6 years and a net present value of $3108 (19). Simultaneously, Rolewicz-Kalinska et al. analyzed the correlation between waste collection and CE strategies, projecting that by 2030, the biogas volume generated from solid waste will increase to nearly 9 million cubic meters annually, resulting in an annual production of renewable energy of almost 17,000 kilowatt-hours (20). The investigation demonstrated that CE strategies have productive impact on solid and hazardous waste generation. CE strategies will maximize product utility by extending the lifespan of products and components, reducing waste generation in both the public and private sectors. Furthermore, CE strategies can reduce the total waste by converting solid waste into biogas for fuel, providing promising prospects for solid waste management.

2.2.5 Land Use and Cover

Land is one of the vital resources that play various essential roles in our daily lives. It can produce food and biomass products, as well as store, transform, and filter substances like carbon, nitrogen, and water. Wiprächtiger et al. assessed the specific

TABLE 2.1
The Specific Impact of the Circular Economy Strategy on Air and Water Quality, Energy Consumption, Natural Resources, Solid and Toxic Waste, and Land Use and Cover Is Different

Strategy Options	Specific Impacts of Applications	References
Energy consumption	Waste plastic fuel pyrolysis was evaluated as an alternative to diesel, resulting in a 4.7% improvement in brake thermal efficiency and a 7.8% reduction in fuel consumption. The organic fraction undergoes anaerobic degradation during the waste treatment process, producing biogas for the power plant. The sorted dry waste can be incinerated at power plants to replace grid power. The non-recoverable high calorific value fraction can be used as waste-derived fuel in cement plants, replacing fossil fuels.	(7)
Natural resources	Circular economy measures can have a positive impact on industries that rely heavily on the continued degradation of natural resources, thereby reducing and optimizing the consumption of these resources. The circular model partially replaces primary raw materials with secondary materials, significantly reducing the consumption of natural resources. Once implemented, the combination of circular economy and bioeconomy can improve the efficiency of natural resource utilization through a systematic approach.	(10, 11)
Air and water quality	Circular economy strategies do not actually improve air quality but prevent it from deteriorating further. Expanding the circular economy through electrification of transportation can reduce PM2.5 emissions by 22% and significantly improve air quality. Circular economy strategies can have significant synergy with gray water reuse as an alternative water source, ensuring that recycled water is safe and meets relevant water quality standards.	(13–15)
Wastes	Product design and business models emphasize multi-functional products, extending the life of products and components, and smart manufacturing to maximize product utility, thereby reducing waste generation in the public and private sectors. Circular economy strategies can reduce overall waste by converting solid waste into biogas for use as fuel.	(24–26)
Land use and cover	Increased use or recycling of bio-based materials, combined with increased use of biological materials, may have more significant negative impacts on land use and land cover. Mitigating climate change by using plants grown on barren land and producing biochar from waste avoids competing with fertile land and food needs. Energy crops grown on marginal lands can produce cellulosic biomass without competing with food crops, aiding in the reclamation of marginal lands, and significantly reducing GHG emissions.	(29, 30)

effects of the CE on land use through a contextual analysis. The results reveal that increasing the use or recycling of bio-based materials and combining with greater utilization of bio-materials might have more significant adverse impacts on land use and land cover. Furthermore, the requirement for larger areas to cultivate bio-based materials puts pressure on land use (21). Fidelis et al. conducted a study on the European Union's CE policies. Land-related CE strategies may put pressure on land use due to the renewable, biodegradable or compostable nature of bio-based materials in the context of the bioeconomy (22). Although the aforementioned studies indicate that CE policies can affect land use, there are solutions when applying the concept of CE to potential land use and land cover. Planting vegetation on marginal land and producing biochar from waste to mitigate climate change can prevent competition with fertile land and food demand. Lepakoski et al. found that biochar produced from willow trees on Finland's marginal land can offset 7.7% of annual agricultural greenhouse gas (GHG) emissions. In addition, cultivating energy crops on marginal land can produce cellulose biomass without competing with food crops, aiding in land reclamation, significantly reducing GHG emissions, or posing no risk to food security (23). Nevertheless, CE strategies remain a significant challenge concerning land use and land cover.

Based on the review, the CE has a generally positive impact on most applications, including reducing energy consumption, recycling resources, and improving air and water quality, among various advantages. However, it still presents significant challenges in terms of land use and cover. This implies that the effectiveness of CE strategies varies depending on the application. Table 2.1 illustrates the specific impacts of CE strategies on various applications and important decisions.

2.3 THE POTENTIAL OF CE IN ACHIEVING NET-ZERO EMISSIONS

In order to keep global warming below the 1.5°C threshold set by the Paris Agreement, all human-generated carbon dioxide (CO_2) emissions must reach net-zero around the mid-21st century. In a net-emission-stable world, carbon emissions become a paramount concern. Any carbon extracted from underground sources must be permanently returned to these sources to prevent eventual release into the atmosphere. Additionally, carbon released into the atmosphere must be recaptured to prevent an increase in atmospheric carbon concentration, which leads to a rise in global average temperatures. This section will explore how the net-zero pathway can be combined with circular approaches and demand-side measures in the agricultural and industrial environments, offering comprehensive solutions. It emphasizes areas that require improvement and the synergies they create, along with broader societal and environmental impacts.

2.3.1 Agricultural Sector

In agricultural development, climate change locate is the most significant factor affecting the industry and serves as a primary driving force (24). Agricultural-related emissions account for 12% of the total GHG emissions, approximately 710 billion tons of carbon dioxide equivalent (Gt CO_{2-eq}) (25). Furthermore, climate change has a severe impact on agricultural productivity. Driven by economic development and population growth, global demand for agricultural products is expected to grow by up to 50% by the middle of the 21st century (26, 27). This could potentially exceed the limits set by international climate goals (28, 29). Therefore, agriculture must advance toward a strategy of net-zero emissions. Previous research has already emphasized the potential of mitigating climate change via improvements in food production and supply chains (30). To effectively attain net-zero emissions in agriculture, it is essential to understand the primary sources of GHG emissions and align them with the goals that can be developed through CE strategies.

2.3.1.1 Current Strategy

The development of agricultural strategies involves complex design and customization, heavily influenced by regional social, economic, environmental, and political factors. Due to the constraints imposed by uncertainties related to actual potential and technological potential, some current policies that combine CE and net-zero emissions are as follows:

(a) Agriculture, fertilizers, and pesticides

Energy use in agriculture involves the use of fossil fuels to operate machinery and electricity for pumping water and irrigation. From the CE perspective, these two aspects can be addressed as follows: transitioning from fossil fuels to renewable fuels for machinery operation and using electricity generated from renewable sources to power agricultural needs. When considering the entire lifecycle assessment of electric and conventional machinery, GHG emissions are reduced by approximately 70%, equating to a yearly reduction of around 720 million tons of $CO_{2\text{-eq}}$ (31). For instance, solar energy is considered one of the most reliable and eco-friendly energy sources globally, with significant potential for application in agriculture. Photovoltaic agriculture is an emerging method aimed at providing green energy for the agricultural sector, with the primary challenge being the sustainable usage of this energy, which has a positive impact on rural environments (32) and contributes to alleviating global energy crises (33). However, as this renewable energy technology becomes more widespread, it may have adverse impacts on land and water resources (34). As suggested by Sampaio and González, it is essential to enhance people's understanding of photovoltaic agriculture, expand research and development of new technologies, and formulate public policy plans to promote the technology (35). Another CE approach is utilizing biogas via anaerobic digestion systems. While this may generate excess resource waste, it can also be transformed into usable resources. Furthermore, there may be the prospect of farm-scale energy farms in the future, producing biogas for use as vehicle fuel and nutrient recovery (36). Organic materials can be redirected to biogas plants to generate clean power. This power can supply energy to both light and heavy vehicles, reducing the use of gasoline/diesel. Once this strategy can change energy usage, reduce potential environmental impacts, and generate income, it becomes highly attractive in the context of the CE (37–39).

Fertilizers and pesticides play a crucial role in increasing global food production in agriculture. The production, distribution, and transportation of mineral fertilizers and pesticides consume approximately 5.4 exajoules (EJ) of energy each year, accounting for 42% of agricultural energy consumption. According to research by Mosquera-Losada et al., the use of organic waste-based fertilizers in the cultivation process can be confirmed by changes in soil fertility indicators of plant components. Additionally, crops obtained using this type of fertilizer yield results similar to traditional methods (40). Therefore, partially or completely replacing traditional fertilizers may be an option in the context of agricultural CE concepts. On the other hand, nitrogen and phosphorus possess a significant position in traditional fertilizers, and phosphorus can be extracted from limestone with the potential for reuse as agricultural fertilizer. However, high concentrations of nitrogen and phosphorus fertilization can lead to nutrient runoff, causing river eutrophication (41, 42). These fertilizer source materials are often scarce and require importation. The use of CE approaches may be a potential solution, such as adsorption of associated phosphorus minerals from soil and waste during ash waste treatment (43). Although there are many scientific publications on phosphorus in these applications, more comprehensive methods are still needed to assess the efficiency of phosphorus in CE methods.

(b) Crop planting

Different crops can use different CE approaches. Adhya et al. proposed that during rice cultivation, flooding conditions lead to an anaerobic soil environment, causing the decomposition of organic matter to produce methane. The flooding prevents oxygen from entering the soil, which creates an environment conducive to methane-producing microorganisms. Flood period reduction has been shown to be an effective technique for reducing methane emissions because it decreases bacterial methane production, thereby lowering methane emissions. In addition, returning rice straw to the fields in the form of biochar avoids burning or anaerobic conditions, thus reducing emissions (44). Moreover, fungal cultivation such as mushrooms, is usually produced from lignocellulosic materials such as straw, sawdust, and wood chips. Daniel and Wösten propose a CE approach to commercial mushroom cultivation. Mushroom fungi convert low-quality waste into compost, as a substrate for other mushroom fungi, as animal feed, to promote animal health, to produce packaging materials, building materials, biofuels, enzymes, etc. (45). These applications could make agricultural production more sustainable and efficient, especially if the CO_2 emissions and heat generated during mushroom cultivation can be used to promote plant growth in greenhouses. Sasha et al. discussed the function of bananas in tropical fruit production. Bananas generate approximately 114.08 million tons of waste every year, leading to environmental problems such as excessive GHG emissions. These wastes can be used to obtain organic fertilizers, bioplastics, or biofuels (ethanol, biogas, hydrogen and biodiesel) via processes for anaerobic or bacterial fermentation. In addition, banana pseudostem waste can be employed to produce nanofibers and silver nanoparticles, helping to reduce waste and enhance the sustainability of agricultural production (46).

(c) Livestock husbandry and breeding technology

In livestock farming, moving towards the goal of reducing methane emissions from ruminant production, farm management and feed quality improvements are first options (47). Various additives incorporated into livestock diets as feed supplements have been mentioned in the literature as reducing methane emissions (48). Andy et al. developed methane inhibitors to reduce methane emissions by up to 40% while increasing body weight without adversely affecting milk yield or composition (49). This provides a way to reduce methane emissions from dairy cows. However, the efficacy of inhibitors and feed additives often decreases significantly over time due to adjustments in the ruminal microbial ecosystem (50). The CE approach provides a potential means of reducing carbon emissions from the livestock industry, while also actively developing biomass energy produced from livestock and poultry manure. The biogas technology has proven to be the most effective method of managing livestock manure from an economic and environmental perspective (51). Yu-nan et al. proposed China's traditional pig industry and a CE model based on biogas. Among them, biogas-based CE models are becoming an effective method for sustainable livestock waste management while minimizing GHG impacts (52). However, relying on biogas to replace fossil energy to achieve carbon reduction is still a topic with development potential. Because the price of renewable energy is generally higher than that of traditional energy, policy support is needed to expand its application (53). Previous research has shown that combining carbon-pricing policies with renewable energy subsidies can support carbon emission reductions (54). The formulation of policies related to the substitution of fossil energy by biogas can provide a reference for governments to achieve global environmental benefits and the development of a CE in the livestock industry.

2.3.2 Industrial Sector

In recent years, the chemical industry has been releasing approximately 2 billion metric tons of CO_2 annually, including direct emissions and energy-related emissions, which accounts for about 5% of global GHG emissions. Furthermore, chemical products have deeply integrated into various supply chains, with approximately 96% of manufactured goods involving chemical components (55). Therefore, the chemical industry confronts unique challenges, as most chemical products contain carbon elements, making it difficult to achieve carbon emissions reduction. There are currently several technology pathways for producing net-zero emissions chemicals based on biomass, carbon utilization, recycling, and carbon capture and storage technologies. However, the feasibility of these technological pathways is limited by local energy and natural resources (e.g., land and water resources). While many studies have assessed the feasibility of a net-zero chemical industry with global resource participation, unresolved issues still exist due to geographical variations (56–58).

Therefore, different countries need to consider local needs and resources when achieving net-zero emissions from the chemical industry, while adjusting technical and environmental conditions.

In this chapter, we specifically focus on geographical diversity and evaluate the feasibility of technological pathways to achieve net-zero emissions in chemical production. The research results indicate that achieving net-zero emissions in the chemical industry requires adopting a comprehensive solution that combines net-zero technology pathways with CE and demand management measures. This comprehensive solution must vary regionally based on available renewable energy, land, and water resources.

2.3.2.1 Available Routes for Net Emission of Chemicals

The chemical industry can continue to provide services while achieving the goal of net-zero emissions through various production pathways. As shown in Figure 2.2, the primary available procedures for chemical production include:

(a) Business as Usual
This plan means maintaining the current way of emitting CO_2. Fossil fuels, the main source of energy required to produce industrial products, contain most of the carbon and hydrogen atoms. However, this fossil fuel-based production process releases large amounts of net CO_2 emissions over the life cycle of the product, which is much shorter than climate change alone. A significant portion of these emissions come from the synthesis of products, especially carbon-based chemicals whose carbon content ultimately ends up in combustion or decomposition. Additionally, leakages in fossil fuel extraction, preparation processes, and supply chains generate other additional CO_2 emissions that may account for approximately 20% of total emissions (59). In this environment, a model called "Business as Usual" is being implemented. This range of measures to improve efficiency and reduce carbon emissions includes waste heat recovery, coal reduction, facility electrification, and a shift from steam boilers to steam/power cogeneration (60, 61). For example, despite the expansion of Europe's chemical industry, its CO_2 emissions fell by 58% from 1990 to 2017 (62). Therefore, further improving chemical production efficiency has a relatively limited effect on energy conservation and emission reduction, and more revolutionary measures need to be taken.

(b) Carbon Capture and Storage (CCS)
This solution achieves net-zero emissions by capturing CO_2 and storing it in appropriate underground geological structures or building materials. In the CCS pathway, most chemicals are still synthesized using fossil fuels and existing organic chemistry technologies, similar to Business as Usual. This capture process can be performed using direct air capture (DAC) or point source capture (PSC) combined with DAC. DAC primarily captures CO_2 from the air, and PSCs are large emitters (such as industrial facilities) equipped with technology to capture CO_2 and transfer it to storage, thereby preventing its emission. In general, PSCs are more cost-effective and energy efficient.

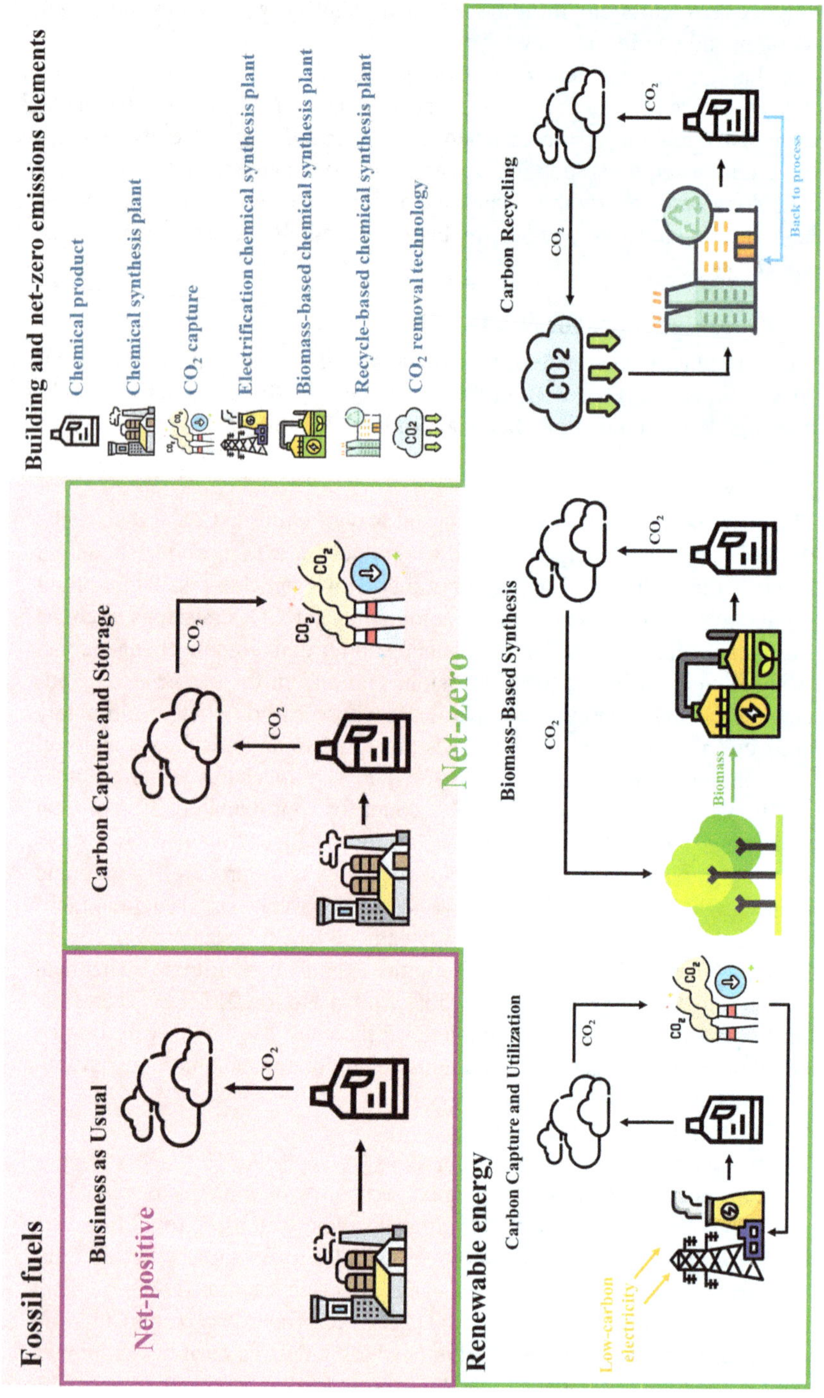

FIGURE 2.2 Achieving net-zero chemical industry in available processes.

Currently, CCS approaches have reached commercial scale, with costs ranging from tens of dollars per ton of CO_2 captured by concentrated sources to hundreds of dollars per ton of direct air capture (63, 64). Although the CCS method has very limited deployment in hard-to-abate industries, it solves the current problem of continued use of fossil fuels and improves large amounts of carbon storage capacity, making a significant contribution to net-zero emissions (65).

(c) Carbon Capture and Utilization (CCU)

This proposition involves capturing carbon and converting it into valuable products or using carbon-free energy via hydrogen electrification to produce chemicals, thereby achieving net-zero emissions. In the CCU approach, the chemical industry achieves net-zero carbon emissions by replacing carbon-based chemicals with carbon sources. The method converts low-energy CO_2 from DAC or PSC into high-energy carbon products. The application of CCU in converting CO_2 into the required carbon-based products requires the development of new organic chemistry technologies, chemical processes, and catalysts (66–68). On the other hand, CO_2 can be converted into syngas (syngas, mainly a mixture of hydrogen, carbon monoxide, and CO_2), methane, methanol or dimethyl ether, and then converted into the required carbon products through other pathways. In all cases, CCU requires low hydrocarbons and energy as inputs for product synthesis since these inputs no longer come from fossil fuels. It is important to note that while permanent CO_2 storage is a unique element of CCS, CO_2 capture is necessary for both CCS and CCU. These CCU technologies have the potential to reduce carbon emissions and reduce reliance on fossil carbon, but still require further development and improvement for wider application in the chemical industry.

(d) Biomass-Based Synthesis

Use biomass as raw material to sustainably synthesize chemicals and achieve net-zero emissions. The hydrogen is conceptually similar to the CCU pathway but is particularly suited to the production of carbon-free chemicals like ammonia. Net-zero emissions of ammonia can be achieved by changing the source of hydrogen atoms, replacing them with renewable energy, and performing water electrolysis (69, 70). The electrification of the chemical industry is the most critical element in this process, as it does not involve carbon and does not require carbon capture. While ammonia can theoretically achieve net-zero emissions by using fossil fuels combined with CCS, CCS infrastructure is more suited to handling organic chemicals that are difficult to decarbonize. In summary, the hydrogen pathway offers a more sustainable way to produce carbon-free chemicals and shift energy toward greener, renewable energy sources, thereby achieving net-zero emissions. This approach has the potential to reduce the carbon emissions of the chemical industry and help address the challenge of climate change.

(e) Carbon Recycling

Reducing emissions by recycling carbon is often combined with other net-zero pathways to offset emissions that cannot be recycled. In reality, biomass contains carbon, hydrogen atoms, and the energy needed to synthesize chemical products. Nonetheless, the chemical structure of biomass feedstocks is often less ideal than fossil fuels, which is often with higher moisture content and lower energy content (71). In the biomass, CO_2 is captured from the atmosphere through photosynthesis and then accumulated during the growth of biomass. These carbon atoms are used to synthesize biomass-based products, and when the product reaches the end of its useful life, some or all of the carbon atoms are released back into the atmosphere, resulting in net-zero CO_2 emissions (72, 73). The biomass pathway helps reduce the chemical industry's carbon emissions because it captures CO_2 from the atmosphere and converts into useful products, while releasing CO_2 at the end of the product's use and life cycle. The renewable nature of biomass resources makes it more environmentally friendly option, helping to mitigate climate change. However, biomass has some disadvantages such as lower energy density and challenges related to demand management. But it remains an important sustainable resource that effectively achieves net-zero carbon emissions goals (74).

All these net-zero production pathways can be combined with demand management strategies to reduce the resources required to achieve net-zero emissions. Importantly, achieving net-zero emissions relies on carbon-free energy sources to ensure the feasibility of these production pathways.

2.4 DIFFERENT STRATEGIES FOR CE

This section discusses strategies for advancing the CE, reviewing past efforts in resource cycling that have already achieved high levels of recycling rates and volumes. However, in the current rapidly changing environment, there is still potential for further enhancement. The concept of CE originated from the period of resource scarcity. The saving and reuse of materials are the driving force for resource recycling (75). With the onset of industrialization (approximately from the 1990s onwards), the system of resource cycling gradually improved due to regulatory developments, and private sector industries began recycling larger quantities of easily recoverable materials, achieving high rates of recycling. Currently, there is a resource revolution underway that still holds significant transformative potential, such as the CE outlined by the Ellen MacArthur Foundation, emphasizing high value, waste reduction, and resource efficiency (76). New transformations facilitated through practices such as exchange, sharing, maintenance, refurbishment, remanufacturing, redesign, and system integration. This approach brings innovative recycling systems, creates more sustainable business opportunities, and reduces dependence on new resources.

To concretely discuss feasible strategies for upgrading to the next generation of the CE, this section analyzes the challenges and opportunities in the development of the CE from three perspectives. Firstly, from an industry perspective, it explores the incentives for industries to embrace the CE and the obstacles that need to be overcome when adopting circular practices. It also discusses which industries may have an advantage and which may lose their foothold if the CE becomes a mainstream trend. Secondly, it examines whether the consumer market can support the demand for CE industries. It delves into the incentives and barriers for consumers to choose CE products or services, including economic and cultural factors. Current consumption modes may need systematic and phased adjustments to steer the market towards the CE. Finally, the section focuses on several innovative technologies and business tendencies, with the potential to promote circular production and consumption systems, leading to disruptive breakthroughs in the future. It emphasizes the need for stimulating research and development in these disruptive innovations, fostering the commercialization of these new industries, and even gaining a competitive advantage in the international market.

2.4.1 The Industrial Perspective on CE

Industries pursue the creation of commercial benefit, thus the concept of the CE can be applied to develop new business models that are promoted by corporate innovation, helping customers obtain more value, and economic growth with fewer resources (77, 78). The ways in which the CE concept creates value can be categorized as follows:

1. Durable Resources: Used resources can be continuously recycled and regenerated over time, which not only extends the service life of the product, but also ensures the sustainable use of resources. Prioritizing the use of sustainable renewable resources in production can break the conflict between resource scarcity and economic development needs, such as the use of renewable energy and biochemical materials (79, 80).
2. High Liquidity in Markets: Optimize utilization efficiency to ensure the convenience of accessing and exchanging products and assets among users (81). By eliminating the idle time of products on the market and increasing the cumulative number of users, usage frequency, or usage time of the product, higher resource efficiency can be achieved with the same number of products.
3. Extended Product Lifecycle: Extend the life of products in the economy, meet more demands, and provide greater performance without requiring additional resources. Practices such as product serviceability, reprocessing, upgrading for maintenance, and warranty extension can extend the product's useful life, creating additional value by providing services (82).
4. Integration of Industries: Connect various wastes to industry in a recycling and connection manner, convert wastes into usable resources, and incorporate them into new production processes (83). Replace downgrading recycling through general recycling and upcycling, which is achieved by reducing resource value degradation, thereby realizing the vision of a waste-to-resource CE across multiple supply chains. Practices include increasing recycling rates, improving recycling classification quality, and efficient resource utilization.

Five circular business models are show in Figure 2.3, combined with other innovative technologies. Enable enterprises to not only improve efficiency, but also establish

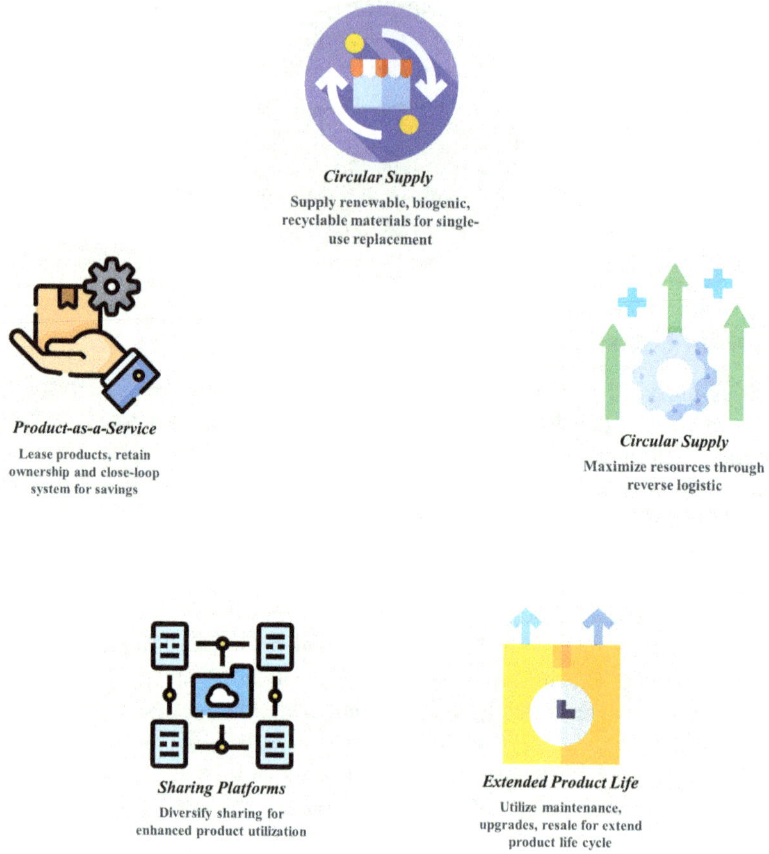
Circular Supply
Supply renewable, biogenic, recyclable materials for single-use replacement
Product-as-a-Service
Lease products, retain ownership and close-loop system for savings
Circular Supply
Maximize resources through reverse logistic
Sharing Platforms
Diversify sharing for enhanced product utilization
Extended Product Life
Utilize maintenance, upgrades, resale for extend product life cycle

FIGURE 2.3 Five business models of circularity.

circular advantages through customer-centric business. These five circular business models are:

1. Circular Supply: Providing materials based on renewable energy, biogenic renewable materials, or entirely recyclable materials to replace single-use consumables.
2. Enhanced Veins: Maximize the use of valuable resources by linking resources of discarded products or by-products through reverse logistics.
3. Extended Product Life: Utilize practices such as maintenance, upgrades, and resale. Products and components will be reused and the life cycle of products within the economy will be extended.
4. Sharing Platforms: Develop a variety of ways to share utilization, use, and ownership on information and trading platforms to improve product utilization.

5. Product-as-a-Service: Enterprises provide the right to use products while retaining ownership, thereby achieving the development of a fully closed-loop system within the manufacturing enterprise and saving raw material and service revenue.

These business models have their unique characteristics, and companies can choose to adopt them individually or in combination. Companies can use models to increase resource productivity while enhancing differentiation, maintaining customer value, reducing costs, creating new revenue streams, and mitigating risk. This transformation requires a new mindset at the executive level, along with a comprehensive set of capabilities that encompass strategy, technology, and operations. Enterprises shift their focus from the profit model of production growth to the participation in continuous product services. They should build collaborative CE networks by fully coordinating the efforts of suppliers, manufacturers, retailers, service providers, and customers.

2.4.2 The Consumer Perspective on the CE

Consumer behavior still mainly follows the framework of classical economics, which represents consumers choosing the best product combinations within a fixed budget based on rationality. In other words, keywords such as circular, green, and sustainable are assumed to provide no utility or value to consumers. Consumers will not pay higher prices or accept lower quality, materials, or waste reduction for products or services that incorporate concepts such as green design, use of recycled materials, etc. (84, 85). Therefore, when considering the promotion of the CE from a consumer perspective, it can be discussed at two positions. First, in cases where concepts like resource sustainability and recycling do not enhance consumer utility, how can products or services use the concept of the CE to create commercial value? This is to convince consumers that, even if they do not consider the environmental impact reduction value of circular products or services, these products or services still have economic incentives for consumers based on their intrinsic value (86, 87). On the other hand, from the perspective of reducing environmental impact, concepts such as resource sustainability and recycling do create value and utility for society. However, this environmental value is often not adequately considered in regular economic activities. In the field of environmental economics, when the value of environmental resources is misvalued by the market, it is called environmental externality (88, 89). The incentives for consumers to purchase CE products and services, without considering the environmental external benefits, can be discussed from several points:

1. Purchasing "Services" Instead of "Products": For infrequently used or high-cost products, adopting a business model that provides services instead of selling traditional products can be economically advantageous for consumers (90). This approach allows consumers to avoid the high initial investment when purchasing the product for the first time and enjoy convenient services such as maintenance, upkeep, and upgrades continuously. From a CE perspective, sales and services allow companies to ensure that the entire

life cycle of a product from cradle to grave is under the control of the manufacturer. This provides companies with economic incentives to extend the product's lifespan through improved design and maintenance, thus increasing resource efficiency.

2. Sharing Economy: Homeowners can benefit from renting out these infrequently used items, effectively reducing the cost of owning them (91). This is economically advantageous for consumers and can activate idle assets. Platforms like ride-sharing services on the internet, which match drivers and riders, not only provide consumers with tangible economic benefits by carpooling but also significantly reduce per capita emissions from transportation. Additionally, reduced environmental impacts due to reduced traffic volume and improved road service levels can be substantial.
3. Other New Business Models: During the transition from traditional linear economic models to circular economic models, traditional industries are confronted with the challenge, but new business models will also emerge. For example, the combination of the Internet of Things (IoT) and autonomous vehicles could completely reshape the future transportation industry. While regular consumers may not actively participate in the development of these new business models, they automatically become advocates if the new models can create more significant benefits with fewer resources. By internalizing the environmental external benefits of circular products and services into the market mechanism, consumers will consider these benefits in their decision-making process, further enhancing their motivation to purchase CE products and services.

2.4.3 Disruptive Innovation Perspective

In the past, resource recycling seemed to have reached obstacles. In recent years, there has been a lack of significant breakthroughs in resource recovery. Many types of materials still cannot be effectively recycled domestically, and some recyclable materials with low quality and low value are circulated on the market or exported to international markets. The linear consumption, where products are disposed of after use, remains the mainstream in the market and has been a significant reason for the low growth of the CE in the past (92).

While many countries have established comprehensive recycling systems, there are still many products with complex material assemblies that are not separated and recycled after product disposal or are only reused at low value. The global economic development model of the past, characterized by a linear "Take-Make-Dispose" approach, has led to many products not achieving high utilization rates after purchase (93). In addition, the cost threshold of resource recycling technology as raw materials is relatively high, making it difficult for companies to establish recycling systems for certain materials. The current consumer culture has a low acceptance of products that promote resource recycling or reuse, which poses a challenge to the development of the CE.

Faced with these issues, the CE emphasizes not only "Reuse," "Remanufacture," and "Recycle" but also "Rethink." Innovative technologies and new business models will change the current linear production and consumption, creating higher utility and value for society through resource reuse (94, 95). Several disruptive innovation trends have been identified in reports by the Ellen MacArthur Foundation and McKinsey, presenting opportunities for breakthroughs in the CE. At the Disruptive Innovation Festival in 2015, industry representatives and experts from around the world discussed the potential impact of disruptive innovations on the future of the CE and overcom current obstacles (96).

Many disruptive innovations, such as (a) 3D printing (97), (b) sharing economy (98), (c) big data (99), (d) modular manufacturing (100), (e) virtual (101), and (f) bionics and nanotechnology (102), will make disruptive breaks to the future development of CE. Future industrial innovation policies should continue to be promoted around these directions.

2.5 CONCLUSIONS

This section reviews the development of the CE, explores the relationship between net-zero emissions and the CE, and discusses different strategies to make corresponding responses. The article begins by analyzing the concept of the CE, including an introduction to the 5Rs (Reduce, Reuse, Recycle, Recover, and Rethink). Furthermore, CE strategies have various impacts on energy consumption, natural resources, air and water quality, waste management, and land use and cover. Specifically, CE strategies can reduce energy and natural resource consumption, improve air and water quality, and effectively manage solid and hazardous waste. However, the use of bio-based materials in CE strategies may increase pressure on land use and land cover.

Additionally, we emphasize the potential for achieving net-zero emissions in the agricultural and industrial sectors through various technologies, innovations, and practices and their integration with CE strategies. Although achieving net-zero emissions through more mature technologies and practices (commercial or pilot-phase solutions) is feasible, their adoption is constrained by underinvestment, a lack of effective policies, and high costs, especially in developing countries. To overcome these challenges, it is crucial to remove adoption barriers and implement strategies that promote the widespread adoption of sustainable practices. Further research is needed to make these technologies economically viable and scalable while understanding their socio-environmental impacts. Similarly, we emphasize the importance of interlinking the pathways to net-zero emissions and combining them with circular and demand-side measures. Further research is needed to determine the optimal combinations of net-zero pathways to minimize resource scarcity caused by decarbonizing agriculture and industry.

Finally, we discuss new ways of thinking and strategies for upgrading the CE transformation. From a disruptive innovation perspective, it seems that the recycling rates of past resources have reached a bottleneck. In recent years, there have been limited breakthroughs in resource recovery, with many types of materials still unable to be effectively recycled domestically. Although many countries have established comprehensive recycling systems, there are still many products with complex material components that are not separated and recycled after disposal, or they are only re-used at low value. Additionally, the current consumer culture has a lower acceptance of products that promote resource recycling or reuse, posing a challenge to the development of the CE. Faced with these bottlenecks in the CE, the CE emphasizes not only "reuse," "remake," and "recycle" but also "rethink." Innovative technologies combined with new business models will provide opportunities to change the current linear production and consumption patterns, creating higher benefits and value for society through resource utilization. These ideas provide a blueprint for future research and development of CE policies and technologies, giving investors, consumers, researchers, and policy regulatory agencies a broader understanding of CE models.

REFERENCES

[1] Hegab, H., Khanna, N., Monib, N., Salem, A., 2023. Design for sustainable additive manufacturing: A review. *Sustainable Materials and Technologies* **35** e00576.
[2] Cudečka-Puriņa, N., Atstāja, D., Koval, V., Purviņš, M., Nesenenko, P., Tkach, O., 2022. Achievement of sustainable development goals through the implementation of circular economy and developing regional cooperation. *Energies* **15** 4072.
[3] Milousi, M., Souliotis, M., 2023. A circular economy approach to residential solar thermal systems. *Renewable Energy* **207** 242–252.
[4] Wang, T., Zhang, M., Springer, C. H., Yang, C., 2021. How to promote industrial park recycling transformation in China: An analytic framework based on critical material flow. *Environmental Impact Assessment Review* **87** 106550.
[5] Moriguchi, Y., 2007. Material flow indicators to measure progress toward a sound material-cycle society. *Journal of Material Cycles and Waste Management* **9** 112–120.
[6] Laskurain-Iturbe, I., Arana-Landín, G., Landeta-Manzano, B., Uriarte-Gallastegi, N., 2021. Exploring the influence of industry 4.0 technologies on the circular economy. *Journal of Cleaner Production* **321** 128944.
[7] Padmanabhan, S., Giridharan, K., Stalin, B., Kumaran, S., Kavimani, V., Nagaprasad, N., Tesfaye Jule, L., Krishnaraj, R., 2022. Energy recovery of waste plastics into diesel fuel with ethanol and ethoxy ethyl acetate additives on circular economy strategy. *Scientific Reports* **12** 5330.
[8] Yaro, N. S. A., Sutanto, M. H., Baloo, L., Habib, N. Z., Usman, A., Yousafzai, A. K., Ahmad, A., Birniwa, A. H., Jagaba, A. H., Noor, A., 2023. A comprehensive overview of the utilization of recycled waste materials and technologies in asphalt pavements: Towards environmental and sustainable low-carbon roads. *Processes* **11** 2095.
[9] Apostu, S. A., Gigauri, I., Panait, M., Martin-Cervantes, P. A., 2023. Is Europe on the way to sustainable development? Compatibility of green environment, economic growth, and circular economy issues. *International Journal of Environmental Research and Public Health* **20** 1078.
[10] Chiappetta Jabbour, C. J., De Camargo Fiorini, P., Wong, C. W. Y., Jugend, D., Lopes De Sousa Jabbour, A. B., Roman Pais Seles, B. M., Paula Pinheiro, M. A., Ribeiro da Silva, H. M., 2020. First-mover firms in the transition towards the sharing economy in metallic natural resource-intensive industries: Implications for the circular economy and emerging industry 4.0 technologies. *Resources Policy* **66** 101596.
[11] Rahman, M. M., Khan, I., Field, D. L., Techato, K., Alameh, K., 2022. Powering agriculture: Present status, future potential, and challenges of renewable energy applications. *Renewable Energy* **188** 731–749.
[12] Abad-Segura, E., Batlles-DelaFuente, A., González-Zamar, M.-D., Belmonte-Ureña, L. J., 2021. Implications for sustainability of the joint application of bioeconomy and circular economy: A worldwide trend study. *Sustainability* **13** 7182.

[13] Su, C., Urban, F., 2021. Circular economy for clean energy transitions: A new opportunity under the COVID-19 pandemic. *Applied Energy* **289** 116666.
[14] Sgroi, M., Vagliasindi, F. G. A., Roccaro, P., 2018. Feasibility, sustainability and circular economy concepts in water reuse. *Current Opinion in Environmental Science & Health* **2** 20–25.
[15] Singh, A. K., Pal, P., Rathore, S. S., Sahoo, U. K., Sarangi, P. K., Prus, P., Dziekański, P., 2023. Sustainable Utilization of Biowaste Resources for Biogas Production to Meet Rural Bioenergy Requirements. *Energies* **16** 5409.
[16] Lal, A., Renaldy, T., Breuning, L., Hamacher, T., You, F., 2023. Electrifying light commercial vehicles for last-mile deliveries: Environmental and economic perspectives. *Journal of Cleaner Production* **416** 137933.
[17] Xiong, X., Liu, X., Yu, I. K. M., Wang, L., Zhou, J., Sun, X., Rinklebe, J., Shaheen, S. M., Ok, Y. S., Lin, Z., Tsang, D. C. W., 2019. Potentially toxic elements in solid waste streams: Fate and management approaches. *Environmental Pollution* **253** 680–707.
[18] Sharma, H. B., Vanapalli, K. R., Samal, B., Cheela, V. R. S., Dubey, B. K., Bhattacharya, J., 2021. Circular economy approach in solid waste management system to achieve UN-SDGs: Solutions for post-COVID recovery. *Science of the Total Environment* **800** 149605.
[19] Al-Wahaibi, A., Osman, A. I., Al-Muhtaseb, A. H., Alqaisi, O., Baawain, M., Fawzy, S., Rooney, D. W., 2020. Techno-economic evaluation of biogas production from food waste via anaerobic digestion. *Scientific Reports* **10** 15719.
[20] Rolewicz-Kalińska, A., Lelicińska-Serafin, K., Manczarski, P., 2020. The Circular Economy and Organic Fraction of Municipal Solid Waste Recycling Strategies. *Energies* **13** 4366.
[21] Wiprächtiger, M., Haupt, M., Heeren, N., Waser, E., Hellweg, S., 2020. A framework for sustainable and circular system design: Development and application on thermal insulation materials. *Resources, Conservation and Recycling* **154** 104631.
[22] Fidélis, T., Cardoso, A. S., Riazi, F., Miranda, A. C., Abrantes, J., Teles, F., Roebeling, P. C., 2021. Policy narratives of circular economy in the EU – Assessing the embeddedness of water and land in national action plans. *Journal of Cleaner Production* **288** 125685.
[23] Leppäkoski, L., Marttila, M. P., Uusitalo, V., Levänen, J., Halonen, V., Mikkilä, M. H., 2021. Assessing the carbon footprint of biochar from willow grown on marginal lands in Finland. *Sustainability* **13** 10097.
[24] Zhang, X., Yao, G., Vishwakarma, S., Dalin, C., Komarek, A. M., Kanter, D. R., Davis, K. F., Pfeifer, K., Zhao, J., Zou, T., D'Odorico, P., Folberth, C., Rodriguez, F. G., Fanzo, J., Rosa, L., Dennison, W., Musumba, M., Heyman, A., Davidson, E. A., 2021. Quantitative assessment of agricultural sustainability reveals divergent priorities among nations. *One Earth* **4** 1262–1277.
[25] Ortiz-Bobea, A., Ault, T. R., Carrillo, C. M., Chambers, R. G., Lobell, D. B., 2021. Anthropogenic climate change has slowed global agricultural productivity growth. *Nature Climate Change* **11** 306–312.
[26] Beltran-Peña, A., Rosa, L., D'Odorico, P., 2020. Global food self-sufficiency in the 21st century under sustainable intensification of agriculture. *Environmental Research Letters* **15** 095004.
[27] van Dijk, M., Morley, T., Rau, M. L., Saghai, Y., 2021. A meta-analysis of projected global food demand and population at risk of hunger for the period 2010–2050. *Nature Food* **2** 494–501.
[28] Ivanovich, C. C., Sun, T., Gordon, D. R., Ocko, I. B., 2023. Future warming from global food consumption. *Nature Climate Change* **13** 297–302.

[29] Clark, Michael A., Domingo, Nina G. G., Colgan, Kimberly, Thakrar, Sumil K., Tilman, David, Lynch, John, Azevedo, Inês L., Hill, Jason D., 2020. Global food system emissions could preclude achieving the 1.5° and 2°C climate change targets. *Science* **370** 705–708.

[30] Rosenzweig, C., Mbow, C., Barioni, L. G., Benton, T. G., Herrero, M., Krishnapillai, M., Liwenga, E. T., Pradhan, P., Rivera-Ferre, M. G., Sapkota, T., Tubiello, F. N., Xu, Y., Mencos Contreras, E., Portugal-Pereira, J., 2020. Climate change responses benefit from a global food system approach. *Nature Food* **1** 94–97.

[31] Farokhi Soofi, A., Saeed D. Manshadi, Saucedo, A., 2022. Farm electrification: A road-map to decarbonize the agriculture sector. *The Electricity Journal* **35** 107076.

[32] Xue, J., 2017. Photovoltaic agriculture - New opportunity for photovoltaic applications in China. *Renewable and Sustainable Energy Reviews* **73** 1–9.

[33] Bey, M., Hamidat, A., Benyoucef, B., Nacer, T., 2016. Viability study of the use of grid connected photovoltaic system in agriculture: Case of Algerian dairy farms. *Renewable and Sustainable Energy Reviews* **63** 333–345.

[34] Ravi, S., Macknick, J., Lobell, D., Field, C., Ganesan, K., Jain, R., Elchinger, M., Stoltenberg, B., 2016. Colocation opportunities for large solar infrastructures and agriculture in drylands. *Applied Energy* **165** 383–392.

[35] Li, G., Jin, Y., Akram, M. W., Chen, X., 2017. Research and current status of the solar photovoltaic water pumping system – A review. *Renewable and Sustainable Energy Reviews* **79** 440–458.

[36] Vieira, S., Barros, M. V., Sydney, A. C. N., Piekarski, C. M., de Francisco, A. C., Vandenberghe, L. P. S., Sydney, E. B., 2020. Sustainability of sugarcane lignocellulosic biomass pretreatment for the production of bioethanol. *Bioresource Technology* **299** 122635.

[37] Lee, D.-H., 2017. Econometric assessment of bioenergy development. *International Journal of Hydrogen Energy* **42** 27701–27717.

[38] Salvador, R., Barros, M. V., Rosário, J. G. D. P. D., Piekarski, C. M., da Luz, L. M., de Francisco, A. C., 2019. Life cycle assessment of electricity from biogas: A systematic literature review. *Environmental Progress & Sustainable Energy* **38** 13133.

[39] Vega-Quezada, C., Blanco, M., Romero, H., 2017. Synergies between agriculture and bioenergy in Latin American countries: A circular economy strategy for bioenergy production in Ecuador. *New Biotechnology* **39** 81–89.

[40] Mosquera-Losada, M. R., Amador-García, A., Rigueiro-Rodríguez, A., Ferreiro-Domínguez, N., 2019. Circular economy: Using lime stabilized bio-waste based fertilisers to improve soil fertility in acidic grasslands. *Catena* **179** 119–128.

[41] Wurtsbaugh, W. A., Paerl, H. W., Dodds, W. K., 2019. Nutrients, eutrophication and harmful algal blooms along the freshwater to marine continuum. *WIREs Water* **6** e1373.

[42] Papangelou, A., Achten, W. M. J., Mathijs, E., 2020. Phosphorus and energy flows through the food system of Brussels Capital Region. *Resources, Conservation and Recycling* **156** 104687.

[43] Carricondo Anton, J. M., Oliver-Villanueva, J. V., Turégano Pastor, J. V., Raigón Jiménez, M. D., González Romero, J. A., Mengual Cuquerella, J., 2020. Reduction of phosphorous from wastewater through adsorption processes reusing wood and straw ash produced in bioenergy facilities. *Water, Air, & Soil Pollution* **231** 128.

[44] Das, S., Adhya, T. K., 2014. Effect of combine application of organic manure and inorganic fertilizer on methane and nitrous oxide emissions from a tropical flooded soil planted to rice. *Geoderma* **213** 185–192.

[45] Grimm, D., Wosten, H. A. B., 2018. Mushroom cultivation in the circular economy. *Applied Microbiology and Biotechnology* **102** 7795–7803.

[46] Alzate Acevedo, S., Diaz Carrillo, A. J., Florez-Lopez, E., Grande-Tovar, C. D., 2021. Recovery of banana waste-loss from production and processing: A contribution to a circular economy. *Molecules* **26** 5282.

[47] Ibidhi, R., Calsamiglia, S., 2020. Carbon footprint assessment of spanish dairy cattle farms: Effectiveness of dietary and farm management practices as a mitigation strategy. *Animals (Basel)* **10** 2083.

[48] Honan, M., Feng, X., Tricarico, J. M., Kebreab, E., 2021. Feed additives as a strategic approach to reduce enteric methane production in cattle: modes of action, effectiveness and safety. *Animal Production Science* **62** 1303–1317.

[49] Reisinger, A., Clark, H., Cowie, A. L., Emmet-Booth, J., Gonzalez Fischer, C., Herrero, M., Howden, M., Leahy, S., 2021. How necessary and feasible are reductions of methane emissions from livestock to support stringent temperature goals? *Philosophical Transactions of the Royal Society A: Mathematical, Physical and Engineering Sciences* **379** 20200452.

[50] Tan, P., Liu, H., Zhao, J., Gu, X., Wei, X., Zhang, X., Ma, N., Johnston, L. J., Bai, Y., Zhang, W., Nie, C., Ma, X., 2021. Amino acids metabolism by rumen microorganisms: Nutrition and ecology strategies to reduce nitrogen emissions from the inside to the outside. *Science of the Total Environment* **800** 149596.

[51] Awasthi, S. K., Kumar, M., Sarsaiya, S., Ahluwalia, V., Chen, H., Kaur, G., Sirohi, R., Sindhu, R., Binod, P., Pandey, A., Rathour, R., Kumar, S., Singh, L., Zhang, Z., Taherzadeh, M. J., Awasthi, M. K., 2022. Multi-criteria research lines on livestock manure biorefinery development towards a circular economy: From the perspective of a life cycle assessment and business models strategies. *Journal of Cleaner Production* **341** 130862.

[52] Xue, Y.-N., Luan, W.-X., Wang, H., Yang, Y.-J., 2019. Environmental and economic benefits of carbon emission reduction in animal husbandry via the circular economy: Case study of pig farming in Liaoning, China. *Journal of Cleaner Production* **238** 117968.

[53] Yi, Z., Xin-Gang, Z., Yu-Zhuo, Z., Ying, Z., 2019. From feed-in tariff to renewable portfolio standards: An evolutionary game theory perspective. *Journal of Cleaner Production* **213** 1274–1289.

[54] Mostafaeipour, A., Bidokhti, A., Fakhrzad, M.-B., Sadegheih, A., Zare Mehrjerdi, Y., 2022. A new model for the use of renewable electricity to reduce carbon dioxide emissions. *Energy* **238** 121602.

[55] Gabrielli, P., Rosa, L., Gazzani, M., Meys, R., Bardow, A., Mazzotti, M., Sansavini, G., 2023. Net-zero emissions chemical industry in a world of limited resources. *One Earth* **6** 682–704.

[56] D'Angelo, S. C., Cobo, S., Tulus, V., Nabera, A., Martín, A. J., Pérez-Ramírez, J., Guillén-Gosálbez, G., 2021. Planetary Boundaries Analysis of Low-Carbon Ammonia Production Routes. *ACS Sustainable Chemistry & Engineering* **9** 9740–9749.

[57] Bachmann, M., Zibunas, C., Hartmann, J., Tulus, V., Suh, S., Guillén-Gosálbez, G., Bardow, A., 2023. Towards circular plastics within planetary boundaries. *Nature Sustainability* **6** 599–610.

[58] Ioannou, I., Galan-Martin, A., Perez-Ramirez, J., Guillen-Gosalbez, G., 2023. Trade-offs between sustainable development goals in carbon capture and utilisation. *Energy & Environmental Science* **16** 113–124.

[59] Cabernard, L., Pfister, S., Oberschelp, C., Hellweg, S., 2021. Growing environmental footprint of plastics driven by coal combustion. *Nature Sustainability* **5** 139–148.

[60] Kabeyi, M. J. B., Olanrewaju, O. A., 2022. Cogeneration potential of an operating diesel engine power plant. *Energy Reports* **8** 744–754.

[61] Sterkhov, K. V., Khokhlov, D. A., Zaichenko, M. N., Pleshanov, K. A., 2021. A zero carbon emission CCGT power plant and an existing steam power station modernization scheme. *Energy* **237** 121570.

[62] Hauser, P., Heinrichs, H. U., Gillessen, B., Müller, T., 2018. Implications of diversification strategies in the European natural gas market for the German energy system. *Energy* **151** 442–454.

[63] Sovacool, B. K., Baum, C. M., Low, S., Roberts, C., Steinhauser, J., 2022. Climate policy for a net-zero future: ten recommendations for Direct Air Capture. *Environmental Research Letters* **17** 074014.

[64] Rosa, L., Mazzotti, M., 2022. Potential for hydrogen production from sustainable biomass with carbon capture and storage. *Renewable and Sustainable Energy Reviews* **157** 112123.

[65] Paltsev, S., Morris, J., Kheshgi, H., Herzog, H., 2021. Hard-to-Abate sectors: The role of industrial carbon capture and storage (CCS) in emission mitigation. *Applied Energy* **300** 117322.

[66] Peres, C. B., Resende, P. M. R., Nunes, L. J. R., Morais, L. C. D., 2022. Advances in carbon capture and use (CCU) technologies: A comprehensive review and CO_2 mitigation potential analysis. *Clean Technologies* **4** 1193–1207.

[67] Centi, G., Perathoner, S., 2020. Chemistry and energy beyond fossil fuels. A perspective view on the role of syngas from waste sources. *Catalysis Today* **342** 4–12.

[68] Huo, J., Wang, Z., Oberschelp, C., Guillen-Gosalbez, G., Hellweg, S., 2023. Net-zero transition of the global chemical industry with CO(2)-feedstock by 2050: Feasible yet challenging. *Green Chemistry* **25** 415–430.

[69] Wang, Y., Tian, Y., Pan, S. Y., Snyder, S. W., 2022. Catalytic processes to accelerate decarbonization in a net-zero carbon world. *ChemSusChem* **15** e202201290.

[70] Anika, O. C., Nnabuife, S. G., Bello, A., Okoroafor, E. R., Kuang, B., Villa, R., 2022. Prospects of low and zero-carbon renewable fuels in 1.5-degree net zero emission actualisation by 2050: A critical review. *Carbon Capture Science & Technology* **5** 100072.

[71] Negi, S., Jaswal, G., Dass, K., Mazumder, K., Elumalai, S., Roy, J. K., 2020. Torrefaction: A sustainable method for transforming of agri-wastes to high energy density solids (biocoal). *Reviews in Environmental Science and Bio/Technology* **19** 463–488.

[72] Priyadharsini, P., Nirmala, N., Dawn, S. S., Baskaran, A., SundarRajan, P., Gopinath, K. P., Arun, J., 2022. Genetic improvement of microalgae for enhanced carbon dioxide sequestration and enriched biomass productivity: Review on CO_2 bio-fixation pathways modifications. *Algal Research* **66** 102810.

[73] Gabrielli, P., Gazzani, M., Mazzotti, M., 2020. The role of carbon capture and utilization, carbon capture and storage, and biomass to enable a net-zero-CO_2 emissions chemical industry. *Industrial & Engineering Chemistry Research* **59** 7033–7045.

[74] Ahmed, A., Ge, T., Peng, J., Yan, W.-C., Tee, B. T., You, S., 2022. Assessment of the renewable energy generation towards net-zero energy buildings: A review. *Energy and Buildings* **256** 111755.

[75] Tseng, M. L., Ha, H. M., Tran, T. P. T., Bui, T. D., Chen, C. C., Lin, C. W., 2022. Building a data-driven circular supply chain hierarchical structure: Resource recovery implementation drives circular business strategy. *Business Strategy and the Environment* **31** 2082–2106.

[76] Kirchherr, J., Yang, N.-H. N., Schulze-Spüntrup, F., Heerink, M. J., Hartley, K., 2023. Conceptualizing the circular economy (revisited): An analysis of 221 definitions. *Resources, Conservation and Recycling* **194** 107001.

[77] Ferasso, M., Beliaeva, T., Kraus, S., Clauss, T., Ribeiro-Soriano, D., 2020. Circular economy business models: The state of research and avenues ahead. *Business Strategy and the Environment* **29** 3006–3024.

[78] Surya, B., Menne, F., Sabhan, H., Suriani, S., Abubakar, H., Idris, M., 2021. Economic growth, increasing productivity of SMEs, and open innovation. *Journal of Open Innovation: Technology, Market, and Complexity* **7** 20.

[79] Sonu, Rani, G. M., Pathania, D., Abhimanyu, Umapathi, R., Rustagi, S., Huh, Y. S., Gupta, V. K., Kaushik, A., Chaudhary, V., 2023. Agro-waste to sustainable energy: A green strategy of converting agricultural waste to nano-enabled energy applications. *Science of the Total Environment* **875** 162667.

[80] Tarazona, N. A., Machatschek, R., Balcucho, J., Castro-Mayorga, J. L., Saldarriaga, J. F., Lendlein, A., 2022. Opportunities and challenges for integrating the development of sustainable polymer materials within an international circular (bio)economy concept. *MRS Energy & Sustainability* **9** 28–34.

[81] Khalek, S. A., Chakraborty, A., 2023. Access or collaboration? A typology of sharing economy. *Technological Forecasting and Social Change* **186** 122121.

[82] Chen, T. L., Kim, H., Pan, S. Y., Tseng, P. C., Lin, Y. P., Chiang, P. C., 2020. Implementation of green chemistry principles in circular economy system towards sustainable development goals: Challenges and perspectives. *Science of the Total Environment* **716** 136998.

[83] Kurniawan, T. A., Dzarfan Othman, M. H., Hwang, G. H., Gikas, P., 2022. Unlocking digital technologies for waste recycling in Industry 4.0 era: A transformation towards a digitalization-based circular economy in Indonesia. *Journal of Cleaner Production* **357** 131911.

[84] Moshood, T. D., Nawanir, G., Mahmud, F., Mohamad, F., Ahmad, M. H., AbdulGhani, A., Kumar, S., 2022. Green product innovation: A means towards achieving global sustainable product within biodegradable plastic industry. *Journal of Cleaner Production* **363** 132506.

[85] Wang, J. X., Burke, H., Zhang, A., 2022. Overcoming barriers to circular product design. *International Journal of Production Economics* **243** 108346.

[86] Han, J., Heshmati, A., Rashidghalam, M., 2020. Circular economy business models with a focus on servitization. *Sustainability* **12** 8799.

[87] Pieroni, Marina P. P., McAloone, Tim C., Pigosso, Daniela C. A., 2019. Configuring New Business Models for Circular Economy through Product–Service Systems. *Sustainability* **11** 3727.

[88] Mohan, S. V., Katakojwala, R., 2021. The circular chemistry conceptual framework: A way forward to sustainability in industry 4.0. *Current Opinion in Green and Sustainable Chemistry* **28** 100434.

[89] Chaudhuri, A., Subramanian, N., Dora, M., 2022. Circular economy and digital capabilities of SMEs for providing value to customers: Combined resource-based view and ambidexterity perspective. *Journal of Business Research* **142** 32–44.

[90] Kanatlı, M. A., Karaer, Ö., 2022. Servitization as an alternative business model and its implications on product durability, profitability & environmental impact. *European Journal of Operational Research* **301** 546–560.

[91] Filippas, A., Horton, J. J., Zeckhauser, R. J., 2020. Owning, using, and renting: Some simple economics of the "sharing economy". *Management Science* **66** 4152–4172.

[92] Patwa, N., Sivarajah, U., Seetharaman, A., Sarkar, S., Maiti, K., Hingorani, K., 2021. Towards a circular economy: An emerging economies context. *Journal of Business Research* **122** 725–735.

[93] Scheel, C., Aguiñaga, E., Bello, B., 2020. Decoupling economic development from the consumption of finite resources using circular economy. A Model for Developing Countries. *Sustainability* **12** 1291.

[94] Donner, M., Gohier, R., de Vries, H., 2020. A new circular business model typology for creating value from agro-waste. *Science of the Total Environment* **716** 137065.

[95] Hankammer, S., Brenk, S., Fabry, H., Nordemann, A., Piller, F. T., 2019. Towards circular business models: Identifying consumer needs based on the jobs-to-be-done theory. *Journal of Cleaner Production* **231** 341–358.

[96] Trento, L. R., Pereira, G. M., Jabbour, C. J. C., Ndubisi, N. O., Mani, V., Hingley, M., Borchardt, M., Gustavo, J. U., de Souza, M., 2021. Industry-retail symbiosis: What we should know to reduce perishable processed food disposal for a wider circular economy. *Journal of Cleaner Production* **318** 128622.

[97] Zhu, C., Li, T., Mohideen, M. M., Hu, P., Gupta, R., Ramakrishna, S., Liu, Y., 2021. Realization of circular economy of 3D printed plastics: A review. *Polymers (Basel)* **13** 744.

[98] Henry, M., Schraven, D., Bocken, N., Frenken, K., Hekkert, M., Kirchherr, J., 2021. The battle of the buzzwords: A comparative review of the circular economy and the sharing economy concepts. *Environmental Innovation and Societal Transitions* **38** 1–21.

[99] Awan, U., Shamim, S., Khan, Z., Zia, N. U., Shariq, S. M., Khan, M. N., 2021. Big data analytics capability and decision-making: The role of data-driven insight on circular economy performance. *Technological Forecasting and Social Change* **168** 120766.

[100] Farrell, C. C., Osman, A. I., Doherty, R., Saad, M., Zhang, X., Murphy, A., Harrison, J., Vennard, A. S. M., Kumaravel, V., Al-Muhtaseb, A. H., Rooney, D. W., 2020. Technical challenges and opportunities in realising a circular economy for waste photovoltaic modules. *Renewable and Sustainable Energy Reviews* **128** 109911.

[101] Rocca, R., Rosa, P., Sassanelli, C., Fumagalli, L., Terzi, S., 2020. Integrating virtual reality and digital twin in circular economy practices: A laboratory application case. *Sustainability* **12** 2286.

[102] Schiros, T. N., Mosher, C. Z., Zhu, Y., Bina, T., Gomez, V., Lee, C. L., Lu, H. H., Obermeyer, A. C., 2021. Bioengineering textiles across scales for a sustainable circular economy. *Chem* **7** 2913–2926.

3 Advancements in Carbon Capture Technologies

Wei-Sheng Chen and Cheng-Han Lee
National Cheng Kung University, Tainan City, Taiwan

3.1 INTRODUCTION: BACKGROUND OF CO_2 CAPTURE

3.1.1 CO_2 AND CLIMATE CHANGE

CO_2 (Carbon dioxide) is a vital component of the carbon cycle and plays a crucial role in maintaining the Earth's climate and supporting life on our planet. CO_2 can provide a greenhouse effect on the Earth, and the greenhouse effect is a natural phenomenon whereby certain gases such as CO_2, H_2O, O_3, CH_4, and so on present in the Earth's atmosphere trap the heat from the Sun [1]. This process plays an important role in maintaining a livable and comfortable environment by preventing excessive heat from escaping into space. CO_2 in the atmosphere is mainly released through both natural and human activities. Natural sources of CO_2 include volcanic eruptions, respiration by plants and animals, and the decay of organic matter. However, the primary driver of increased CO_2 levels in the atmosphere is human activity, particularly burning fossil fuels such as coal, oil, and natural gas [2].

The Industrial Age, which began in the 18th century, brought about significant changes in human activities, particularly in industrialization and the burning of fossil fuels. These activities have released large amounts of CO_2 into the atmosphere, contributing to the phenomenon known as anthropogenic or human-caused climate change [3]. Before the Industrial Age, the concentration of CO_2 in the atmosphere was relatively stable at around 280 ppm Parts per million for several thousand years [4]. However, due to the burning of fossil fuels, as well as deforestation and other land-use changes, the concentration of CO_2 has been increasing. According to NOAA ESRL Global Monitoring Laboratory, the global atmospheric CO_2 concentration had reached 424 ppm [5], about 51% higher than pre-industrial levels, continuing a steady climb further into situation not seen for millions of years.

As the concentration of CO_2 rises, it increases the Earth's average temperature and causes climate change [6, 7]. Here are some key impacts of CO_2 on climate change:

1. Increase of Global Temperature: As CO_2 and other greenhouse gases accumulate, they trap more heat, causing the Earth's temperature to rise. The temperature rise has numerous consequences, including melting ice caps, rising sea levels, and altered weather patterns.

DOI: 10.1201/9781003463009-3

2. Ocean Acidification: When CO_2 dissolves in seawater, it forms carbonic acid, leading to ocean acidification. This process lowers the pH of the oceans, which can have detrimental effects on marine life, particularly organisms that rely on calcium carbonate for their shells or skeletons, such as coral reefs and shellfish. Ocean acidification can disrupt marine ecosystems and have cascading effects throughout the food chain.
3. Altered Weather Patterns: Increasing CO_2 levels can change weather patterns, such as shifts in precipitation and more frequent and intense extreme weather events. These changes can result in more frequent droughts, heatwaves, heavy rainfall events, and stronger storms, impacting ecosystems, agriculture, and human communities [8].
4. Melting of Ice Caps and Glaciers: Rising temperatures caused by CO_2 emissions contribute to melting ice caps and glaciers. This phenomenon leads to the loss of freshwater resources, rising sea levels, and increased vulnerability to coastal flooding in many regions. It also affects the habitats of various species, disrupts ecosystems, and threatens human settlements in low-lying areas.
5. Ecosystem Disruption: Climate change driven by CO_2 emissions can disrupt ecosystems and biodiversity. Many species struggle to adapt to rapid shifts in temperature, altered rainfall patterns, and changing habitats. This can lead to species extinction, species distribution changes, and ecological community imbalances.

3.1.2 Net-Zero Emissions

Due to the effect of CO_2 concentration in the atmosphere on climate change, achieving net-zero emissions has become a paramount goal for countries, organizations, and individuals worldwide by 2050. Net-zero emissions refer to the balance between the number of GHGs (greenhouse gases) emitted into the atmosphere and the amount removed or offset. This balance ensures that the overall impact on the Earth's climate is neutral, effectively halting further influence of global warming. To achieve net-zero emissions by 2050, several key strategies and measures are implemented, such as decarbonizing energy systems, electrifying transportation, afforestation and reforestation, industrial decarbonization, and technological innovations. Among them, Energy Transition and CCUS (Carbon capture, utilization, and storage) are the major ways to achieve the goal. Take Taiwan as an example, Taiwan's net-zero emission plan is mainly expected to enhance the usage of renewable energy during 2020–2030 and explore CCUS techniques during 2030–2050 (Figure 3.1) [9, 10]. As a novel technique, CCUS is a set of technologies and processes to reduce CO_2 emissions from industrial and energy-related sources and prevent them from entering the atmosphere [11]. During the CCUS process, CO_2 can be utilized in various ways instead of being released into the atmosphere after CO_2 capture [12–15]. Carbon utilization involves converting captured CO_2 into useful products or feedstocks. For instance, CO_2 can be applied to chemicals, fuels, building materials, or even for enhancing oil recovery from depleted oil fields. On the other hand, captured CO_2 can also be stored in underground geological formations. The CO_2 is transported through pipelines or ships and injected deep underground into suitable geological formations, such as depleted oil and gas reservoirs, saline aquifers, or coal seams.

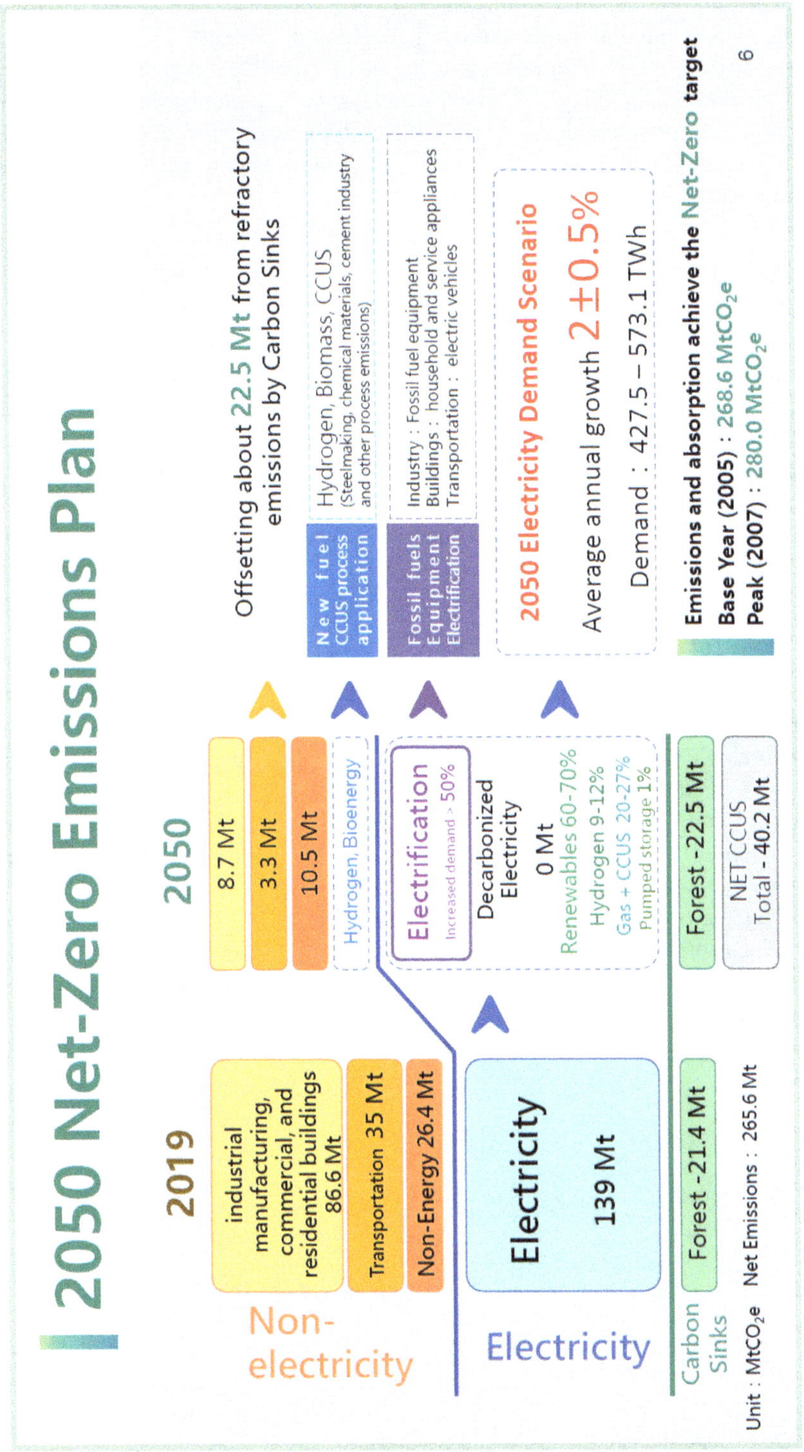

FIGURE 3.1 Taiwan's net-zero emission plan [9, 10].

In the future, we can already foresee that the techniques of CCU and CCS will gradually mature and mitigate climate change. However, whether CCU or CCS process, it needs to capture CO_2 first. As a result, CO_2 capture procedures should be initially established in every industry. Nowadays, CO_2 capture techniques can be classified into three categories, namely Pre-Combustion Capture, Post-Combustion Capture, and Oxy-Fuel Combustion Capture. The following content will demonstrate their principles and characteristics.

3.2 PRE-COMBUSTION CAPTURE

3.2.1 Introduction of Pre-Combustion Capture

Pre-combustion capture refers to removing CO_2 from fossil fuels before combustion is completed. In this approach, the reactor doesn't fully burn the fossil fuels; instead, fossil fuels are partially oxidized in steam and oxygen/air under high temperatures and pressure to form synthesis gas. This synthesis gas is a mixture of H_2, CO, CO_2, and smaller amounts of other gaseous components, such as methane. Afterward, a water-gas shift process is employed to convert CO and H_2O to H_2 and CO_2 [16]. The CO_2 concentration in this mixture has a range of 15–50%. At last, the CO_2 can be captured, separated, transported, and ultimately sequestered, while the fuel rich in H_2 is combusted (Figure 3.2).

Today, the pre-combustion capture of CO_2 is primarily utilized in coal-fired IGCC (Integrated Gasification Combined Cycle) plants. During the process, CO_2 is separated from a gas mixture before combustion takes place. Unlike IGCC plants without CO_2 capture, IGCC plants implementing pre-combustion capture employ a series of components, including a water-gas shift reactor, an acid gas removal unit, a CO_2 compression unit, and a gas turbine powered by hydrogen [17] (Figure 3.3). Shijaz et al. compared power generation methods using no carbon capture and pre-combustion carbon capture. The findings revealed that pre-combustion carbon capture reduces the overall efficiency of power plants compared to plants without CO_2 capture. The capture of CO_2 from the fuel stream decreases the fuel volume directed to the turbine, consequently diminishing power generation [16, 18]. According to the investigation of Mukherjee et al., the energy penalties were 7.1% when conducting

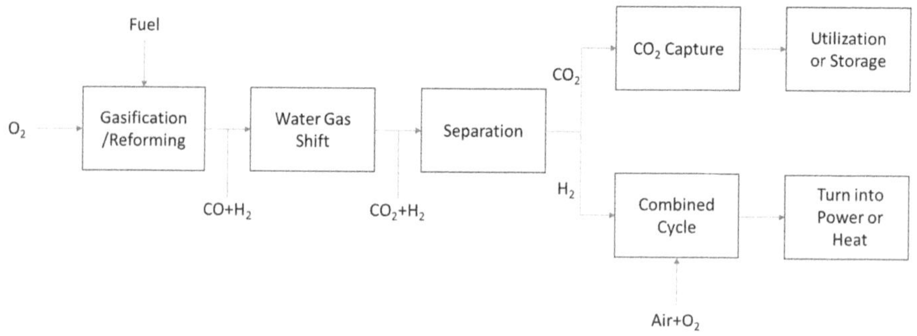

FIGURE 3.2 Flowchart of pre-combustion capture [16].

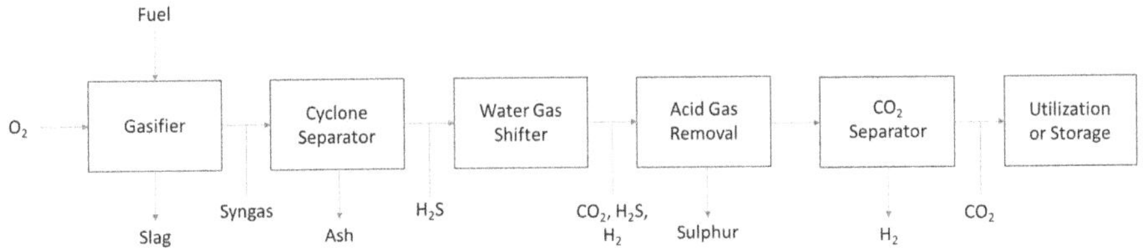

FIGURE 3.3 Scheme of IGCC with pre-combustion capture [16].

the pre-combustion process [16, 19]. Nevertheless, incorporating a CO_2 capture unit becomes imperative considering the environmental implications.

Pre-combustion carbon capture employs a combination of physical and chemical techniques to capture CO_2 from processed syngas. In industrial applications, chemical absorbents like carbonates and physical solvents such as polypropylene glycol and methanol are commonly used for CO_2 capture. While an efficient solvent or absorbent in pre-combustion carbon capture technology can achieve over 90% CO_2 capture, it also reduces plant efficiency [16, 20]. The calcium looping process is an alternative approach for pre-combustion carbon, which proves to be both efficient and cost-effective. In this method, CO_2 capture is accomplished by effectively sorbing CaO, followed by the optimal temperature desorption of $CaCO_3$ to release CO_2. This cyclic process is repeated multiple times, utilizing waste heat from the gasifier to minimize energy consumption in the CO_2 capture process. The pre-combustion carbon capture method using CaO demonstrates high effectiveness, reducing costs and energy consumption [16, 21].

3.2.2 Cases of Pre-Combustion Capture

Several commercial-sized IGCC projects with pre-combustion carbon capture already/under construction. For an IGCC plant with capture based on current technology, the TRL of the major components of the air separation, gasification, gas cooling, shift, sulphur removal, and CO_2 capture are almost at TRL 9. The Kemper County IGCC (also called Kemper County energy facility or Plant Ratcliffe), with a net capacity of 524 MW, is specifically engineered to convert locally sourced lignite into syngas, which is then utilized to power a combined cycle power plant. Additionally, it incorporates the capability to capture CO_2 for the purpose of enhancing oil recovery. The Kemper County IGCC is a 2x1 configuration with two gasification trains, two combustion turbines, two heat recovery steam generators, and a common steam turbine. The process begins with the delivery of lignite to each gasification train from the adjacent mine. The lignite is dried, milled, and subsequently fed to the gasifier, producing syngas. The syngas passes through several cleanup steps to remove particulate matter, ammonia, sulphur species, and CO_2 before fueling the combustion turbine to generate electricity. Electricity is generated through a steam turbine generator using steam produced in the process. Apart from electricity, the plant also yields various marketable products. To reduce carbon emissions to a level similar to that of a natural-gas-fired combined cycle power plant, the plant is designed to capture and sell around 65 percent of the carbon input for enhanced oil recovery (EOR). Additionally, the plant produces roughly 21,000 tons per year of commercial-grade anhydrous ammonia and 169,000 tons per year of sulfuric acid, both of which have market value [22]. An additional instance is the Tianjin IGCC demonstration project, with a capacity of approximately 250 MW. The demonstration project is situated in Lingang Industry Park which is in the central of Tianjin Binhai New Area. The project incorporates a 2000 tons/d TPRI gasifier, a set of 46,000 Nm^3/h Air Separation Unit responsible for generating oxygen at a purity level of 99.5%. Additionally, it utilizes cyclones and ceramic filters to eliminate dust, implements MDEA process and LO-CAT (a patented liquid redox system that uses a proprietary chelated iron solution to convert H_2S to innocuous, elemental sulphur) technology to decrease SO_2 emissions to 1.4 mg/Nm^3, and produce 23 tons of sulphur annually. Also, Tianjin IGCC shows the first step for China's utility companies to reduce CO_2 which is also very important. Utilizing an integrated system of technologies, this process effectively separates and captures CO_2 from the emissions of the industrial plant. The captured CO_2 is subsequently compressed, transported to a geological site, and injected into suitable underground formations at high pressure for cryogenic storage. Although this process can reduce CO_2 emissions, more than two thirds of the CCS cost derivates from CO_2 capture. Thus, it is critical to decrease capture cost [23].

3.3 POST-COMBUSTION CAPTURE

3.3.1 Introduction of Post-Combustion Capture

Post-Combustion Capture is a technology designed to capture CO_2 emissions from the flue gases of power plants and industrial facilities after the combustion process has taken place. The process of post-combustion capture involves separating CO_2 from the flue gases emitted during the combustion of fossil fuels such as coal, natural gas, or oil [24, 25]. These flue gases typically contain a mixture of gases, including nitrogen, water vapor, and pollutants, along with a relatively small concentration of CO_2 (Figure 3.4).

There are several techniques employed in post-combustion capture. One commonly used method is absorption, where a solvent, such as an amine-based solution, is used to selectively absorb the CO_2 from the flue gases. This process typically occurs in a large absorption tower, where the solvent is brought into contact with the flue gases. Once the CO_2 is absorbed, it can be separated from the solvent and stored

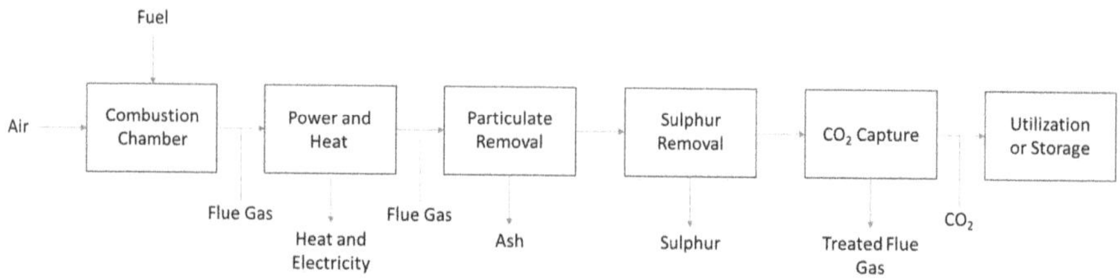

FIGURE 3.4 Flowchart of post-combustion capture [16].

or utilized for various purposes, such as enhanced oil recovery or industrial applications. Commercially available chemical absorption technologies primarily rely on amines, particularly monoethanolamine (MEA), due to their stable operation, high reactivity, and excellent absorption capacity [26–28]. Nonetheless, the MEA process faces significant drawbacks, including a substantial energy requirement for solvent regeneration and amine degradation caused by the high temperatures (around 120 °C) in the regeneration section. Consequently, these drawbacks render the process unappealing for large-scale implementation. In contrast, aqueous ammonia offers various advantages over MEA [29, 30]. It possesses a high CO_2 loading capacity, eliminates absorbent degradation concerns, and operates at a lower regeneration temperature (85–95 °C). This lower temperature allows for the utilization of low-grade thermal energy, enhancing overall energy efficiency. Additionally, aqueous ammonia demonstrates the potential to simultaneously capture CO_2, SO_2, and NO_x from flue gas. However, it is essential to incorporate an ammonia abatement system into the aqueous ammonia process since ammonia can volatilize under operating conditions, contributing to machine damage or personal injury. Therefore, in addition to MEA and ammonia, other amines such as 3-amino-1-proponal, 5-amino-1-pentanol, diethylamine, and so on have been investigated for carbon capture [31].

The other approach is adsorption, which involves using solid materials called adsorbents to capture CO_2. To meet the necessary adsorption capacities and produce CO_2 captures of adequate purity, it is crucial for materials to exhibit high selectivity for CO_2 compared to other constituents in flue gas, including oxygen, nitrogen, and water. As a result, the emphasis on material development has predominantly centered around creating materials with a strong surface affinity for CO_2. Besides, this method is often used in conjunction with PSA (pressure swing adsorption) or TSA (temperature swing adsorption) processes, where the adsorbent is regenerated by altering the pressure or temperature conditions, releasing the captured CO_2 for further use or storage [32].

Membrane is also commonly used in a post-combustion carbon capture system. Membrane techniques for post-combustion carbon capture involve the use of selectively permeable membranes to separate CO_2 from flue gases emitted during the combustion of fossil fuels [33]. These membranes allow for the selective transport of CO_2 while impeding the passage of other gases. Currently, different types of membranes are used in post-combustion carbon capture, including polymeric and inorganic membranes. Polymeric membranes are made of organic polymers and offer advantages such as low cost, flexibility, and ease of fabrication. These membranes operate based on the principle of solution-diffusion, where CO_2 molecules dissolve in the membrane and then diffuse through it [34, 35]. Inorganic membranes, on the other hand, are typically made of ceramic materials, metals, or composites. These membranes often possess superior thermal and chemical stability compared to polymeric membranes, making them suitable for high-temperature applications. Inorganic membranes can operate based on different mechanisms, such as molecular sieving or surface diffusion, depending on the specific membrane design and material [36, 37]. The main advantage of membrane-based post-combustion carbon capture is its potential for energy efficiency. Membranes can operate at near-ambient temperatures and pressures, reducing the need for energy-intensive regeneration processes compared to traditional solvent-based capture methods. Additionally, membranes can be more easily integrated into existing systems and scaled up. However, there are also challenges of membrane-based carbon capture, such as membrane degradation due to contaminants and temperature conditions [38]. Therefore, achieving high CO_2 selectivity while maintaining sufficient permeability remains an ongoing research focus.

To sum up, post-combustion capture technology has gained significant attention in reducing CO_2 emissions from existing power plants and industrial facilities. It can be retrofitted to existing infrastructure, making it a potentially cost-effective solution for reducing carbon emissions. Nonetheless, the process requires energy for operation, which can impact the overall efficiency of the power plant or facility. Research and development efforts are focused on optimizing the capture process and developing more efficient solvents, adsorbents, or membranes to improve the overall economics of post-combustion capture.

3.3.2 Cases of Post-Combustion Capture

Post-combustion capture of CO_2 in coal-fired power plants is a straightforward approach to mitigating carbon emissions. This method entails installing a facility that captures and eliminates CO_2 from the flue gases emitted by the boiler, preventing their release into the atmosphere. Moreover, this type of carbon capture plant is the most convenient to retrofit into existing coal-fired power plants, which makes it particularly significant in the event of mandatory carbon capture requirements.

In 2014, the Boundary Dam Power Station in Canada became the first power station in the world to successfully use CCS technology. The project uses a post-combustion capture technique that captures 90% of Unit 3's CO_2. The post-combustion capture technique employed at Boundary Dam involves using a solvent, typically an amine-based solution, which absorbs the CO_2 from the flue gas stream. The solvent is then separated from the absorbed CO_2, which is further compressed to a supercritical state for transportation and storage. Besides, the SO_2 emissions from the coal process are also reduced by up to 100% [39, 40]. The CO_2 that is captured is stored underground instead of being released into the atmosphere. The Boundary Dam Integrated CCS Demonstration is the first coal plant with a fully integrated post-combustion capture system. The project has been successful in capturing a total of 625,000 tons of CO_2 since the capture started in 2014.

The Petra Nova project, undertaken by NRG Energy and JX Nippon Oil, is an extensive initiative aimed at reducing carbon emissions from a coal-burning power plant in Thompsons, Texas. This multi-million-dollar project involves retrofitting one of the plant's boilers at the WA Parish Generating Station with a post-combustion carbon capture treatment system. The objective is to capture around 90% of the CO_2 emissions from a 240 MW slipstream of flue gas, effectively treating a significant portion of the atmospheric exhaust. This endeavour aims to annually sequester or utilize approximately 1.4 million metric tons of this greenhouse gas [41]. The captured CO_2 traveled via an 80-mile pipeline to an oilfield near Houston for use in EOR operations to increase extraction. Petra Nova's target CCS capture rate was 90%, and

NRG claims it met the target. The project became operational in January 2017 and was expected to remove up to 1.6 million tons of carbon annually. However, the project faced several technical problems and downtime, which affected its performance. The project was also affected by the drop in oil prices, which made it less profitable. The project was shut down since May 2020 and restarted operations on September 5, 2023 [42].

3.4 OXY-FUEL COMBUSTION CAPTURE

3.4.1 Introduction of Oxy-Fuel Combustion Capture

Oxy-fuel combustion capture involves fuel combustion using nearly pure oxygen (around 98%) to ensure that the resulting flue gas primarily consists of CO_2 and water, with minimal amounts of other gases. This method simplifies the process of recovering CO_2 by condensing it from a purge stream rich in CO_2 and water, while containing very low levels of nitrogen (Figure 3.5). In contrast, conventional combustion involves the combustion of CO_2 in the air, which is predominantly composed of approximately 79.9% nitrogen [43]. The advantage of oxy-fuel combustion capture is that it requires less energy for CO_2 separation and compression steps compared to other carbon capture technologies [44]. Moreover, it also has the characteristics of high CO_2 concentrations and partial pressure (<50% of post-combustion capture), reduced NO_x emissions, high CO_2 purity, lower gas volumes due to increased density, and has proven to improve fuel efficiency, thus saving up to 40% CO_2 emissions [45]. On the other hand, the primary challenge associated with this approach lies in separating oxygen from air, which typically requires cryogenic methods that consume significant amounts of energy. Nevertheless, an encouraging advancement known as chemical looping combustion is currently being developed. This innovative technique involves the extraction of oxygen from the air by oxidizing a metallic compound. During combustion, this compound can be reduced, releasing oxygen. This process holds great potential for achieving complete CO_2 capture.

The integration of Chemical Looping Combustion (CLC) in oxy-fuel combustion capture represents a promising advancement in carbon capture technology. In this approach, CLC facilitates the separation of oxygen from air, addressing one of the main challenges associated with oxy-fuel combustion. Unlike conventional methods that rely on cryogenic separation, CLC employs a unique process where a

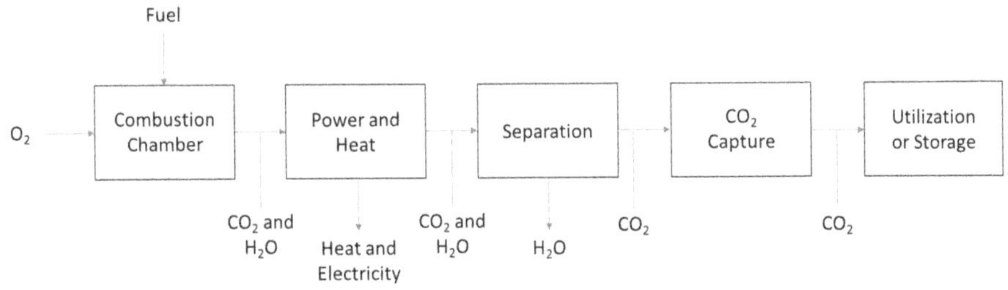

FIGURE 3.5 Flowchart of oxy-fuel combustion capture [16].

metallic compound is oxidized to remove oxygen from the air. During combustion, the oxidized metallic compound is reduced, liberating the captured oxygen for combustion. This innovative mechanism offers several advantages over traditional approaches. Firstly, it eliminates the need for cryogenic equipment, reducing energy requirements and associated costs. Additionally, using CLC enables the production of a flue gas composed mainly of CO_2 and water vapor, making the subsequent CO_2 recovery process more efficient. Although CLC is still under development, the application of CLC in oxy-fuel combustion capture represents a promising avenue for advancing carbon capture technologies. It produces oxygen internal to the process, eliminating the large capital, operating, and energy costs associated with oxygen production. [46].

3.4.2 Cases of Oxy-Fuel Combustion Capture

Even if the techniques of Oxy-Fuel combustion capture are still improved, there are some cases of oxy-fuel combustion capture.

1. The Callide Oxy-Fuel Project (Queensland, Australia): The location of the project is the Callide A power station, initially constructed in 1965. Within this power station, one of its 30-MW units underwent recommissioning and transformation, leading to the establishment of the Callide Oxy-Fuel Project. Commencing operations in 2012, the Callide Oxy-Fuel Project ceased its activities in March 2015. Following the discontinuation of the demonstration plant, the project entered its decommissioning phase, which concluded in May 2018, marking the finalization of the project.
2. The White Rose CCS Project (United Kingdom): The White Rose Carbon Capture and Storage project was a plan to construct an oxy-fuel coal-fired power plant adjacent to the Drax power station in the United Kingdom [47]. Proposed in 2012 by Capture Power Limited, in partnership with National Grid, this project aimed to showcase the competitive and environmentally friendly use of oxy-fuel technology in generating low-carbon electricity. With a projected capacity of 426 MW, the plant intended to capture and store 2 Mt of CO_2 annually in an offshore saline aquifer, achieving a capture rate of 90% [47]. However, in April 2016, the project faced a setback when its Development Consent Order application, submitted to the Department of Energy and Climate Change, was rejected. The rejection was primarily due to the project's lack of viable funding, following the UK government's cancellation of a CCS Competition in November 2015.

3.5 COMPARISON AND FUTURE TRENDS OF CO_2 CAPTURE

After introducing three different carbon capture techniques, it can be observed that each has merits and drawbacks. Table 3.1 lists the comparison among three techniques, which are promising ways to mitigate CO_2 emissions [44]. In addition to these three techniques, there are also many different approaches to conducting carbon capture, such as applying algae, desalination brine, and mineral carbonation.

TABLE 3.1
Comparison of Pre-Combustion, Post-Combustion, and Oxy-Fuel Combustion Capture [44]

	Pre-Combustion Capture	Post-Combustion Capture	Oxy-Fuel Combustion Capture
Merit	• Low energy requirements • Syngas can be reused as a fuel	• Most common technology in CO_2 capture methods • Available for existing and new coal-fired power plants	• Produce high efficiency steam cycles • Low pollutants emissions
Drawbacks	• Requires supporting systems • High equipment cost	• High performance and water requirements are needed • Low CO_2 partial pressure at ambient pressure	• High energy consumption • High cost

Microalgae employ efficient photosynthesis strategies, called the CCM (CO_2 concentrating mechanism), to effectively acquire inorganic carbon. The advantages of algae applied for carbon capture are below.

1. Utilizing algae-based carbon capture technology enables the potential sequestration of carbon at a massive gigaton scale. Algae exhibit remarkable efficiency in extracting carbon dioxide from the atmosphere, effectively absorbing an amount of carbon comparable to that absorbed by all land plants and trees combined [48].
2. The application of algae-based carbon capture technology offers the potential to mitigate ocean acidification. Algae have the capability to absorb CO_2 from the atmosphere and transform it into organic matter [49].
3. Algae-based carbon capture technology can cultivate microalgae for carbon capture and utilization. Microalgae are mainly exploited to produce biofuels, fertilizers, and other products [49].

Desalination is a process that removes salt and other minerals from saline water to produce fresh water for human consumption and other uses. However, a by-product of desalination is brine, a highly concentrated saltwater solution that must be disposed of [50]. Disposing of brine can be environmentally challenging as it can have detrimental effects on marine ecosystems if discharged directly into the ocean. One potential solution is to utilize the brine in CCUS systems. The concept involves injecting brine into deep underground geological formations that have suitable characteristics for CO_2 storage. The high salt content of the brine can help enhance the storage capacity and secure the captured CO_2 in the storage reservoir, preventing its leakage over time. Additionally, the interaction between the brine and CO_2 can lead to mineral carbonation, a natural process where CO_2 reacts with minerals such as Na, Mg, and Ca in the brine to form stable carbonates [15, 51]. This mineralization process can turn these metals into carbonate compounds for CCU.

For example, El-Naas M H et al. assessed a novel method for CO_2 capture and reduction of water salinity, inspired by the Solvay process, but without using ammonia. The approach involves reacting CO_2 with saline water, such as reject brine, in the presence of calcium hydroxide. The research investigates the impact of various operational factors like reaction temperature, water pH, and reaction stoichiometry on the efficiency of CO_2 capture and sodium removal for both the conventional and modified Solvay processes. By employing response surface methodology, the optimal conditions for maximizing CO_2 capture efficiency and sodium removal were determined. For both processes, these conditions were found to be at a temperature of 20 °C and a pH greater than 10. Under these optimal conditions, the traditional Solvay process achieved CO_2 capture efficiencies of 86% and sodium removal of 29%, whereas the Modified process achieved higher results with 99% CO_2 capture efficiency and 35% sodium removal. The research identified water pH as a critical parameter influencing the effectiveness of the reaction process, with higher pH values leading to improved performance in both CO_2 capture efficiency and sodium removal. The experimental results indicate that the Modified Solvay process outperforms the traditional approach regarding CO2 capture efficiency, sodium removal, and energy consumption.

(The Solvay process or the ammonia-soda process, is a vital industrial method used to produce sodium carbonate (soda ash) through sodium chloride (common salt) and limestone (calcium carbonate). The primary objective of the Solvay process is twofold: the extraction of sodium carbonate, a key compound used in various industries, and the recovery of ammonia for reuse. The process takes place through a series of chemical reactions that involve the utilization of readily available raw materials. The main chemical reactions involved in the Solvay process are as follows:

$$CaCO_3 \rightarrow CaO + CO_2 \tag{3.1}$$

$$NaCl + NH_3 + CO_2 + H_2O \rightarrow NaHCO_3 + NH_4Cl \tag{3.2}$$

$$2NaHCO_3 \rightarrow Na_2CO_3 + CO_2 + H_2O \tag{3.3}$$

$$CaO + 2NH_4Cl \rightarrow CaCl_2 + 2NH_3 + H_2O \tag{3.4}$$

The modified Solvay process is then using $Ca(OH)_2$ or $Mg(OH)_2$ to react with $NaCl \rightarrow 2NaCl + Ca(OH)_2 + 2CO_2 \rightarrow 2NaHCO_3 + CaCl_2$)

In a nutshell, carbon capture has the potential to play a crucial role in mitigating climate change by significantly reducing CO_2 emissions from major industrial sources. It allows for the continued use of fossil fuels while minimizing their environmental impact. Apart from the above-mentioned capture techniques, many novel methods are still being investigated and improved. For instance, the mineral carbonation of alkaline industrial waste for CO_2 sequestration and utilization will be introduced in the following chapter.

REFERENCES

1. Rodhe H 1990 A comparison of the contribution of various gases to the greenhouse effect *Science* **248** 1217–9.
2. Skripnuk D F and Samylovskaya E A 2018 Human activity and the global temperature of the planet. *IOP Conf. Ser.: Earth. Environ. Sci.* **180** 012021.
3. Ehleringer J R 2005 *A History of Atmospheric CO_2 and Its Effects on Plants, Animals, and Ecosystems (Ecological Studies).* Springer Science & Business Media.
4. Hofmann D J, Butler J H and Tans P P 2009 A new look at atmospheric carbon dioxide *Atmos. Environ.* **43** 2084–6.
5. Broken record: Atmospheric carbon dioxide levels jump again [cited Jun 30 2023] Available from: https://www.noaa.gov/news-release/broken-record-atmospheric-carbon-dioxide-levels-jump-again
6. Lee Z H, Sethupathi S, Lee K T, Bhatia S and Mohamed A R 2013 An overview on global warming in Southeast Asia: CO_2 emission status, efforts done, and barriers *Renew. Sust. Energ. Rev.* **28** 71–81.
7. Rahimpour M R 2020 *Advances in Carbon Capture.* Woodhead Publishing.
8. Jones E, Qadir M, van Vliet M T H, Smakhtin V and Kang S 2019 The state of desalination and brine production: A global outlook *Sci. Total Environ.* **657** 1343–56.
9. National Development Council [cited Jun 30 2023] Taiwan's pathway to net-zero emissions in 2050 Available from: https://www.ndc.gov.tw/en/Content_List.aspx?n=B15472 4D802DC488
10. Chen P-H, Lee C-H, Wu J-Y and Chen W-S 2023 Perspectives on Taiwan's Pathway to Net-Zero Emissions *Sustainability* **15** 5587.
11. Consoli C P and Wildgust N 2017 Current status of global storage resources *Energy Procedia* **114** 4623–8.
12. Li L, Zhao N, Wei W and Sun Y 2013 A review of research progress on CO_2 capture, storage, and utilization in Chinese Academy of Sciences *Fuel* **108** 112–30.
13. Al-Mamoori A, Krishnamurthy A, Rownaghi A A and Rezaei F 2017 Carbon capture and utilization update *Energy Technol.* **5** 834–49.
14. Kang D, Lee M-G, Jo H, Yoo Y, Lee S-Y and Park J 2017 Carbon capture and utilization using industrial wastewater under ambient conditions. *Chem. Eng. J.* **308** 1073–80.
15. Yoo Y, Kang D, Park S and Park J 2020 Carbon utilization based on post-treatment of desalinated reject brine and effect of structural properties of amines for $CaCO_3$ polymorphs control *Desalination* **479** 114325.
16. Madejski P, Chmiel K, Subramanian N and Kuś T 2022 Methods and techniques for CO_2 capture: Review of potential solutions and applications in modern energy technologies *Energies* **15** 887.
17. Sayigh A 2022 *Comprehensive Renewable Energy: Biomass and Biofuel Production.* Elsevier.
18. Shijaz H, Attada Y, Patnaikuni V S, Vooradi R and Anne S B 2017 Analysis of integrated gasification combined cycle power plant incorporating chemical looping combustion for environment-friendly utilization of Indian coal *Energy Convers. Manag.* **151** 414–25.
19. Mukherjee S, Kumar P, Yang A and Fennell P 2015 Energy and exergy analysis of chemical looping combustion technology and comparison with pre-combustion and oxy-fuel combustion technologies for CO_2 capture. *J. Environ. Chem. Eng.* **3** 2104–14.
20. Olabi A G, Obaideen K, Elsaid K, Wilberforce T, Sayed E T, Maghrabie H M, et al. 2022 Assessment of the pre-combustion carbon capture contribution into sustainable development goals SDGs using novel indicators *Renew. Sust. Energ. Rev.* **153** 111710.

21. Wu F, Dellenback P A and Fan M 2019 Highly efficient and stable calcium looping based pre-combustion CO_2 capture for high-purity H_2 production *Mater. Today Energy* **13** 233–8.
22. Nelson M, Rush R, Madden D, Pinkston T and Lunsford L 2012 Kemper County IGCC (tm) Project Preliminary Public Design Report.
23. Asian Development Bank 2016 [cited Jul 4 2023] PRC's tianjin breathing easier with cleaner coal power Available from: https://www.adb.org/results/prc-s-tianjin-breathing-easier-cleaner-coal-power
24. Chao C, Deng Y, Dewil R, Baeyens J and Fan X 2021 Post-combustion carbon capture *Renew. Sust. Energ. Rev.* **138** 110490.
25. Feron P 2016 *Absorption-Based Post-Combustion Capture of Carbon Dioxide*. Woodhead Publishing.
26. Raynal L, Bouillon P-A, Gomez A and Broutin P 2011 From MEA to demixing solvents and future steps, a roadmap for lowering the cost of post-combustion carbon capture *Chem. Eng. J.* **171** 742–52.
27. Luis P 2016 Use of monoethanolamine (MEA) for CO_2 capture in a global scenario: Consequences and alternatives *Desalination* **380** 93–9.
28. Luo X and Wang M 2016 Optimal operation of MEA-based post-combustion carbon capture for natural gas combined cycle power plants under different market conditions *Int. J. Greenh. Gas Control* **48** 312–20.
29. Al-Hamed K H M and Dincer I 2021 A comparative review of potential ammonia-based carbon capture systems *J. Environ. Manag.* **287** 112357.
30. Al-Hamed K H M and Dincer I 2022 Analysis and economic evaluation of a unique carbon capturing system with ammonia for producing ammonium bicarbonate *Energy Convers. Manag.* **252** 115062.
31. Yoo Y, Kang D, Park S and Park J 2020 Carbon utilization based on post-treatment of desalinated reject brine and effect of structural properties of amines for $CaCO_3$ polymorphs control *Desalination* **479** 114325.
32. Aroussi A 2012 *Proceedings of the 3rd International Gas Processing Symposium*. Elsevier.
33. Zamarripa M A, Eslick J C, Matuszewski M S and Miller D C 2018 Multi-objective optimization of membrane-based CO_2 capture *Comput. Aided Chem. Eng.* **44** 1117–22.
34. Laciak D V and Langsam M 2000 *Membrane separations|Gas separations with polymer membranes Encyclopedia of Separation Science*. Elsevier Science Ltd. 1725–38
35. Morreale B 2015 *Novel Materials for Carbon Dioxide Mitigation Technology*. Elsevier.
36. Roberts D L, Abraham L C, Blum Y and Way J D 1992 Gas separation with glass membranes. Final report.
37. Bhave R 2012 *Inorganic Membranes Synthesis, Characteristics and Applications*. Springer Science & Business Media.
38. 王大銘 2019 薄膜科技概論. 五南圖書出版股份有限公司.
39. Boundary Dam Carbon Capture Project [cited Jul 6 2023] Available from: https://www.saskpower.com/Our-Power-Future/Infrastructure-Projects/Carbon-Capture-and-Storage/Boundary-Dam-Carbon-Capture-Project
40. World's First Post-Combustion CCS Coal Unit Online in Canada 2014 [cited Jul 6 2023] Power Magazine Available from: https://www.powermag.com/worlds-first-post-combustion-ccs-coal-unit-online-in-canada/
41. Petra Nova - W.A. Parish Project [cited Jul 5 2023] Available from: https://www.energy.gov/fecm/petra-nova-wa-parish-project
42. Smyth J [cited Jul 5 2023] Petra Nova carbon capture project stalls with cheap oil Available from: https://energyandpolicy.org/petra-nova/

43. Basile A 2011 *Advanced Membrane Science and Technology for Sustainable Energy and Environmental Applications*. Woodhead Publishing.
44. Elhenawy S E M, Khraisheh M, AlMomani F and Walker G 2020 Metal-organic frameworks as a platform for CO_2 capture and chemical processes: Adsorption, membrane separation, catalytic-conversion, and electrochemical reduction of co_2 *Catalysts* **10** 1293.
45. Southwest Research Institute [cited Jul 6 2023] Oxy-Fuel Combustion Available from: https://www.swri.org/industry/advanced-power-systems-conventional-power-generation/oxy-fuel-combustion
46. Mukherjee S, Kumar P, Yang A and Fennell P 2015 Energy and exergy analysis of chemical looping combustion technology and comparison with pre-combustion and oxy-fuel combustion technologies for CO_2 capture *J. Environ. Chem. Eng.* **3** 2104–14.
47. Carbon Capture and Sequestration Technologies @ MIT [cited Jul 6 2023] Available from: https://sequestration.mit.edu/tools/projects/white_rose.html
48. Sustainable Brands [cited Jul 6 2023] Algae may be a 'brilliant' solution for capturing carbon at gigaton scale Available from: https://sustainablebrands.com/read/chemistry-materials-packaging/algae-may-be-a-brilliant-solution-for-capturing-carbon-at-gigaton-scale
49. Singh J and Dhar D W 2019 Overview of carbon capture technology: Microalgal biorefinery concept and state-of-the-art. *Front. Mar. Sci.* **6** 17505-13
50. Lee C-H, Chen P-H and Chen W-S 2021 Recovery of alkaline earth metals from desalination brine for carbon capture and sodium removal *Water* **13** 3463.
51. El-Naas M H, Mohammad A F, Suleiman M I, Al Musharfy M and Al-Marzouqi A H 2017 A new process for the capture of CO_2 and reduction of water salinity *Desalination* **411** 69–75.

4 Mineral Carbonation of Alkaline Industrial Waste and Byproducts for CO_2 Sequestration and Utilization

Hsing-Jung Ho and Atsushi Iizuka
Tohoku University, Sendai City, Japan

4.1 INTRODUCTION

Climate change is a pressing issue, and greenhouse gases (GHG) are the main contributors. Reduction of GHG emissions is the main direction for climate change mitigation. The priority target to achieve carbon neutrality is CO_2, which accounts for approximately three-quarters of GHG emissions.[1]

Reduction of CO_2 emissions cannot be achieved through one method. Instead, a range of methods are required and must be integrated with each other.[2] Various reports have indicated that the energy sector needs to maximize its efforts to develop renewable energy sources and reduce CO_2 emissions from fossil fuels. However, energy transition takes time, while CO_2 emissions from industrial processes are unavoidable. Consequently, carbon capture, utilization, and storage (CCUS) technology is essential. CCUS technology has become increasingly important over time and is expectedly to contribute significantly to carbon neutrality after 2030.[2]

CCUS technology can be divided into two main concepts: permanent storage of captured CO_2 deep underground in geological formations, and the use of captured CO_2 as a raw material in various applications. The former concept is known as carbon capture storage (CCS). CCS is capable of sequestering large amounts of CO_2, and can be combined with other goals, such as enhancement of gas, oil, and coal bed methane recovery, and provision of additional economic benefits.[3,4] Major barriers to the development of CCS technologies are geological sequestration options, economic problems, social acceptance, difficulties in monitoring sequestered CO_2, and possible unknown environmental effects.[5] Another concept that has attracted increasing attention is carbon capture utilization (CCU). Because captured CO_2 is regarded as an alternative raw material, it can be used in a variety of applications while reducing the exploitation of natural resources. For example, it can be used in production of chemicals (e.g., methanol and urea) and plastic, biological conversion, and the food and drink industry. Among the possible applications, conversion of CO_2 into inorganic

 DOI: 10.1201/9781003463009-4

carbonates is a promising option.[5] In this chapter, the use of mineral carbonation technology with industrial waste will be introduced.

4.2 MINERAL CARBONATION TECHNOLOGY

Mineral carbonation technology is one of CCUS technologies that captures CO_2 and converts it into inorganic carbonated minerals for further utilization as shown in Eq. (4.1).

$$\text{Alkaline materials} + CO_2 \rightarrow \text{Carbonated products } (CaCO_3, MgCO_3) \quad (4.1)$$

The carbonation reaction is exothermic and spontaneous, and the CO_2 is stored stably in carbonates. Compared with other CCUS technologies, mineral carbonation is regarded as a more stable route in terms of thermodynamics. Because the carbonation reaction is spontaneous, it occurs in natural environments where CO_2 in the atmosphere reacts with natural silicate minerals, which is known as weathering. However, natural weathering is very slow, usually taking tens or hundreds of millions of years. Consequently, for an application of this reaction to capture CO_2 from industrial waste gas (CO_2 content: 10%–50%),[5] the reaction is expected to be enhanced, also called as accelerated carbonation.

Alkaline materials are used to mineralize CO_2 via various routes. Compared with natural alkaline rocks, alkaline industrial waste and byproducts are generally more reactive. Additionally, this material is generated in huge quantities and is inexpensive. Alkaline waste and byproducts can be used to capture CO_2 via mineral carbonation, which provides simultaneous reduction of CO_2 emissions and utilization of industrial waste. This concept is described in detail in Section 4.2.1.

4.2.1 MINERAL CARBONATION OF INDUSTRIAL WASTE AND BYPRODUCTS

The concept of mineral carbonation using industrial waste and byproducts is shown in Figure 4.1. Industries, such as cement manufacturing, iron-making and steel-making,

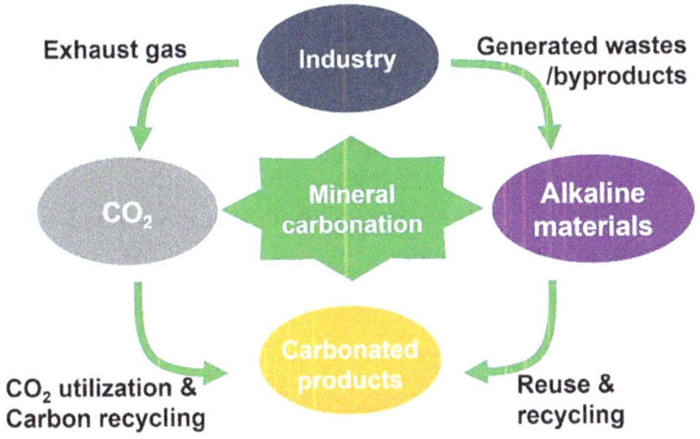

FIGURE 4.1 Concept of mineral carbonation of alkaline waste and byproducts.[17]

coal-fired power generation, incineration, and mining generate large volumes of CO_2 flue gas during the process operation. At the same time, large quantities (~7 Gt) of alkaline waste and byproducts, including cement and concrete wastes,[6–8] iron and steel slags,[9] coal fly ash,[10,11] incineration ash,[12,13] and mining wastes,[14,15] are generated.[16] Therefore, mineral carbonation of industrial waste and byproducts is essential for utilizing CO_2, recycling carbon, and preventing the negative environmental effects of industrial waste disposal. The use of industrial waste and byproducts to capture industrial CO_2 exhaust gas will provide an important pathway for industries to move towards carbon neutrality and a circular economy.

Within this concept, the following key points need to be considered:

1. Composition of industrial waste and byproducts: It is important to understand the composition of waste and byproducts. Although these materials are mainly composed of alkaline Ca and Mg, the main compounds vary according to the production process and have different properties and reactivities.
2. Composition of industrial flue gas: The exhaust gas CO_2 concentration, temperature, and pressure need to be specifically considered. The reaction rate steps and the theoretical CO_2 uptake amount will be affected regarding to the kinetics and thermodynamics. Additionally, the presence of impurities will affect the carbonation reactions.
3. Amounts of flue gases and waste generated: CO_2 flue gases and alkaline waste are the raw materials of this technology; therefore, the development strategy should take into account the amounts of these materials generated.
4. Ultimate goals: In the concept of mineral carbonation of alkaline waste and byproducts, there are two main goals, one is to maximize the CO_2 uptake amount, and the other is to effectively use the waste. According to the purposes, different processes are required to be designed and implemented.
5. Process design: Process design is key for mineral carbonation technology. Effective waste utilization and CO_2 capture is the most important consideration and can involve integration with other operations or issues.
6. Utilization of carbonated products: Depending on the process, carbonated products with different characteristics will be obtained. To increase the value of the overall technology, effective utilization methods must be explored.
7. Calculation methodology of CO_2 emissions reductions: An overall CO_2 emissions reduction methodology must be developed, including various scenarios for different industries and applications of carbonated products.[18] To integrate carbonation technology with the Carbon Border Adjustment Mechanism (CBAM), a holistic calculation approach to CO_2 emissions reduction needs to be developed.

4.2.2 Carbonation Process

Mineral carbonation can be divided into direct carbonation and indirect carbonation. In direct carbonation, CO_2 reacts directly with alkaline waste to form carbonated products in a single stage. Direct carbonation can be further differentiated into direct gas–solid carbonation and direct aqueous carbonation.

Direct gas–solid carbonation: This process is also called direct dry carbonation. Solid compounds containing Ca and Mg react directly with CO_2 gas under pressure, obtaining the carbonated product.[19] Through this method, CO_2 is usually coated on the surface of particles. Conversion of CO_2 into carbonates and diffusion of CO_2 through the product layer are the main rate controlling steps.[20] Provision of a small amount of water as water vapor can enhance the reaction.[21]

Direct aqueous carbonation: In direct aqueous carbonation, there are three main reaction steps: Ca/Mg extraction (Eq. 4.2), CO_2 dissolution (Eq. 4.3), and $CaCO_3/MgCO_3$ precipitation (Eq. 4.4). These steps occur simultaneously and are in competition. The Ca/Mg extraction step is generally facilitated in lower pH environments and will release Ca^{2+}/Mg^{2+} and OH^- simultaneously. The CO_2 dissolution step is favored in alkaline environments. In the CO_2 dissolution reaction, CO_2 dissolves into the aqueous phase to form H_2CO_3, HCO_3^-, and CO_3^{2-}, and H^+ is produced simultaneously. In the $CaCO_3/MgCO_3$ precipitation step, the extracted Ca^{2+}/Mg^{2+} reacts with dissolved CO_3^{2-} to form $CaCO_3/MgCO_3$. The final solution is neutral. The experimental setup is shown in Figure 4.2.[6]

$$Ca(OH)_2/Mg(OH)_2 \rightarrow Ca^{2+}/Mg^{2+} + 2OH^- \tag{4.2}$$

$$CO_2 + H_2O \rightarrow H_2CO_3 \rightarrow H^+ + HCO_3^- \rightarrow 2H^+ + CO_3^{2-} \tag{4.3}$$

$$Ca^{2+} + CO_3^{2-} \rightarrow CaCO_3 \tag{4.4}$$

In addition to direct carbonation, carbonation with multiple steps is called indirect carbonation.[22,23] Indirect carbonation can be performed in various ways, and many

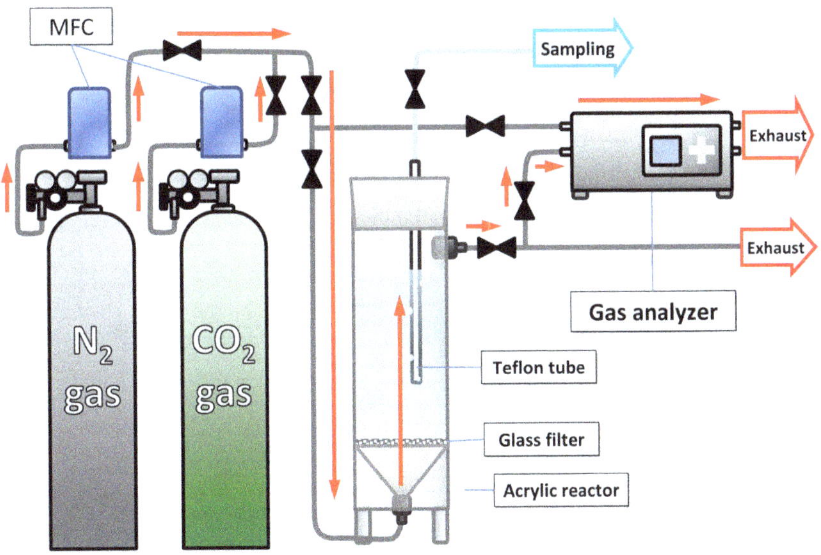

FIGURE 4.2 Schematic of the direct aqueous carbonation setup.[6]

types of carbonated products and byproducts can be obtained. In this section, pH-swing, pressure-swing, and circular indirect carbonation are introduced.

pH-swing carbonation: In the pH-swing method, acidic and alkaline solvents are used to enhance the carbonation reaction.[24] Acidic solvents such as H_2SO_4, HCl, HNO_3, CH_3COOH, and NH_4HSO_4 are used to extract Ca/Mg from the feedstock. Alkaline solvents such as NaOH, NH_4OH, and KOH are used to increase the solution pH to enhance the precipitation of carbonates. The pH-swing method can greatly increase the reaction rate and produce pure carbonated products.[25] Applications for byproducts derived from the extraction step should be developed.[26–28] The feedstock, temperature, pressure, and type of solvents are major parameters that influence the overall carbonation performance.

Pressure-swing carbonation: In pH-swing carbonation, the use of acidic and alkaline solvents increases the process cost and wastewater generation. The use of a pressure-swing avoids these issues. To achieve a pressure-swing, the CO_2 partial pressure is adjusted to control the solubility of $CaCO_3$/$MgCO_3$ and CO_2 gas, which promotes Ca/Mg extraction, CO_2 dissolution, and $CaCO_3$/$MgCO_3$ precipitation. The pressure-swing method occurs in two stages: dissolution and precipitation.[29] Dissolution occurs under a high CO_2 partial pressure, while precipitation occurs under a low CO_2 partial pressure. Because Ca/Mg can be precipitated under low CO_2 partial pressure, the solution can be circulated to the next high CO_2 partial pressure step.

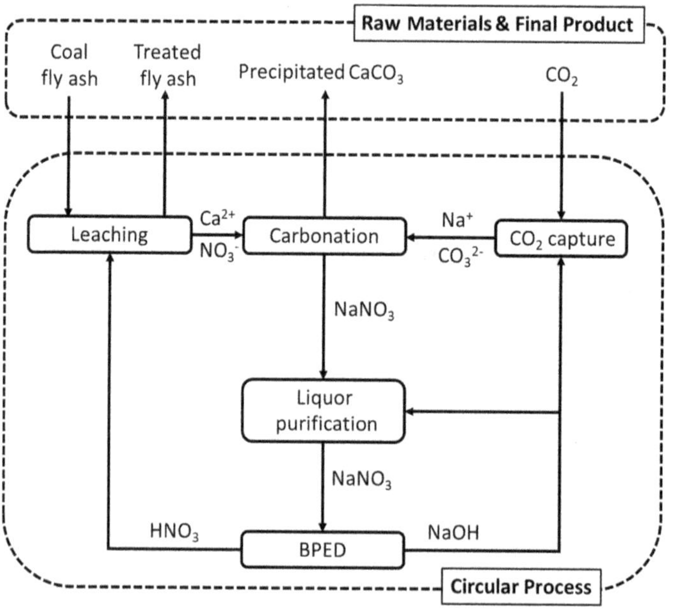

FIGURE 4.3 Proposed process flow diagram of circular indirect carbonation.[11]

Circular indirect carbonation: Circular indirect carbonation (Figure 4.3) is a promising method that was proposed recently.[11,30,31] In this concept, the solution regeneration step is integrated with indirect carbonation to avoid issues with chemical consumption and wastewater generation. The proposed process has four main steps: leaching, CO_2 capture, $CaCO_3$ precipitation, and solution regeneration. In the leaching step, a solution of HNO_3 is used to extract Ca^{2+} from coal fly ash. In the CO_2 capture step, a solution of NaOH is applied to dissolve CO_2 gas in the carbonate solution (i.e., Na_2CO_3). Then, the solution containing the extracted Ca is mixed with a solution of Na_2CO_3 to precipitate high-purity $CaCO_3$. The remaining salt solution (i.e., $NaNO_3$) is treated by bipolar membrane electrodialysis (Eq. 4.5). This regenerates the HNO_3 solution and NaOH solution, which can then be reused in leaching and CO_2 capture, respectively.

$$NaNO_3 + H_2O \rightarrow NaOH\ (\text{for } CO_2 \text{ capture}) + HNO_3\ (\text{for leaching}) \qquad (4.5)$$

4.3 ROLE OF MINERAL CARBONATION IN REDUCING CO_2 EMISSIONS USING ALKALINE INDUSTRIAL WASTE AND BYPRODUCTS

Mineral carbonation is an important method to reduce CO_2 emissions using alkaline industrial waste and byproducts. However, the source and magnitude of CO_2 emissions and the characteristics of the industrial waste will vary by country according to the structure of its energy and industrial sectors. Therefore, understanding the role of mineral carbonation in achieving the goal of carbon neutrality is very important for formulating strategies. In this chapter, the role of mineral carbonation of alkaline waste in reducing CO_2 emissions is presented for a case study in Taiwan.[17]

In Taiwan, CO_2 accounts for more than 95% of GHG emissions.[32] The net CO_2 emissions in 2019 were 252.1 Mt. These emissions are associated with the energy sector and continuing industrial and commercial development. Consequently, reduction of CO_2 emissions will require many changes, including reduction of process-related emissions from industry, fuel combustion, transportation, and commercial activities. Most importantly, CCUS technology is required to manage unavoidable CO_2 emissions such as process-related emissions. Taiwan is in a seismic zone and is densely populated; therefore, the importance of developing CCU is greater than in some other countries. Consequently, the development of mineral carbonation technology is critical. The main goal of the application of mineral carbonation technology to industrial waste is to reduce CO_2 emissions from the industry.

To determine the CO_2 reduction potential, the CO_2 uptake capacity of each waste is required; however, it is difficult to calculate this simply from the content of Ca and Mg because of the different availabilities of different chemical forms of Ca and Mg in alkaline waste and byproducts. The calculation method has been described elsewhere,[33] and CO_2 uptake capacity data can be obtained from various publications. The amount of CO_2 that can be sequestered via mineral carbonation of alkaline waste and byproducts can be calculated using Eq. (4.6). The CO_2 reduction potential for mineral carbonation by using alkaline waste/byproducts for industries at which the waste and byproducts are produced can be calculated using the CO_2 reduction contribution for the industry (Eq. 4.7).

$$CO_2 \text{ reduction potential (Mt } CO_2) = \text{Quantity of alkaline waste/by product generated (Mt alkaline waste/byproduct)} \times CO_2 \text{ uptake capacity } [(\text{Mt } CO_2)/(\text{Mt alkaline waste/byproduct})]. \quad (4.6)$$

$$CO_2 \text{ reduction contribution of an industry (\%)} = CO_2 \text{ reduction potential of the alkaline waste/byproduct generated by the industry (Mt } CO_2)/CO_2 \text{ emissions from the industry (Mt } CO_2) \times 100. \quad (4.7)$$

From the literature,[17] the total CO_2 reduction potential for mineral carbonation of alkaline waste in Taiwan is 4.2 Mt CO_2. The largest contributors to this reduction potential are iron and steel slags (50.3%) and cement and concrete wastes (36.9%) (Figure 4.4). Other wastes, such as incineration ashes, coal combustion ashes, and paper waste, account for approximately 12% (0.49 Mt CO_2) of total reduction potential.

Considering the annual CO_2 emissions from industry and the CO_2 reduction potentials of the alkaline wastes, the CO_2 reduction contributions (%) from these industries can be determined (Figure 4.5). The CO_2 reduction contribution is an important indicator of the suitability and potential of the technology for the industry. Mineral carbonation technology will be promising for the iron and steel industry and cement and concrete industry, with CO_2 reduction contributions of approximately 14.5% and 28.1%, respectively. Additionally, although incineration ash contains many heavy

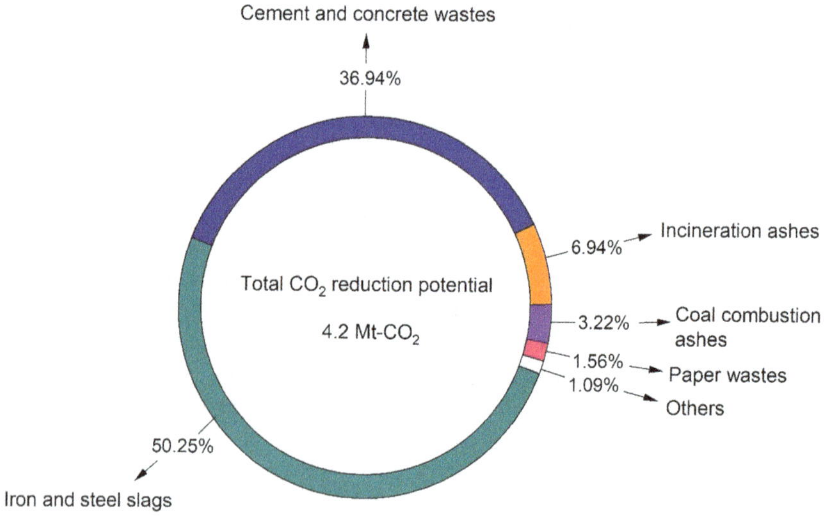

FIGURE 4.4 Percentage contributions to the CO_2 reduction potential via mineral carbonation of alkaline waste and byproducts from various industries in Taiwan.[17]

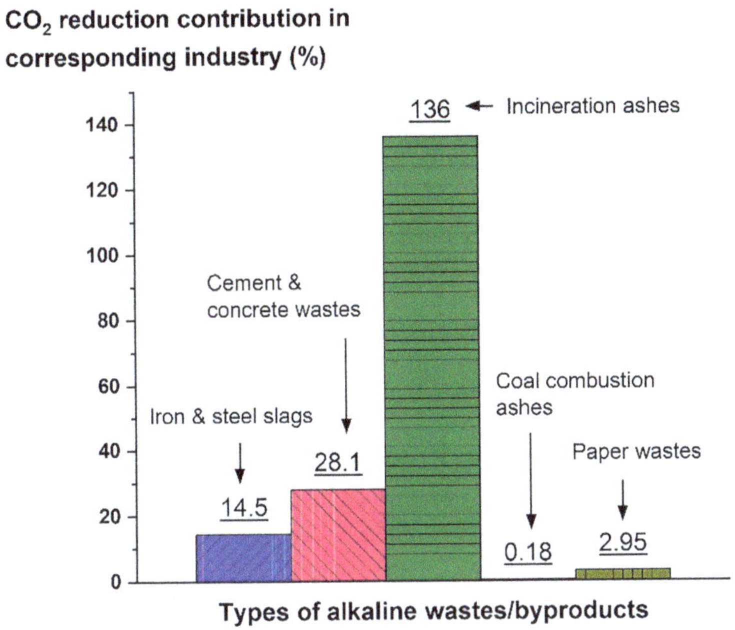

FIGURE 4.5 CO_2 reduction contributions for mineral carbonation using alkaline waste and byproducts from various industries.[17]

metals and impurities, its treatment could offset the CO_2 emitted during incineration. However, the characteristics of incineration ash can vary widely with the source, and the actual CO_2 reduction contribution may be lower than expected. It is worth noting that even though coal combustion ash accounts for 3% of the total CO_2 reduction potential, the CO_2 reduction contribution for coal-fired power generation is less than 0.2%. This difference is attributed to the fact that CO_2 emissions from coal-fired power generation are enormous. The best way to reduce CO_2 emissions from coal-fired power generation is to reduce coal consumption and develop renewable energy.

This preliminary evaluation shows that the iron and steel industry and cement and concrete industry are promising industries to develop mineral carbonation technology to reduce CO_2 emissions in Taiwan. Carbonated products could be used as alternative materials in the cement and concrete industry. Additionally, these products could be developed into materials for soil stabilization, plant growth, and paper fillers.[34] However, the end use of these products should be carefully considered and market analyses of carbonated products must be performed. The structures of the industrial and energy sectors are the main factors that affect development strategy. This technology is also affected by natural resource reserves and policies related to carbon taxation and GHG regulation. More case studies are required to establish a comprehensive roadmap to mineral carbonation technology in different countries and under different conditions.

4.4 CALCULATION OF CO_2 EMISSIONS REDUCTION USING MINERAL CARBONATION TECHNOLOGY

Although CO_2 emissions reduction can be obviously achieved by mineral carbonation technology, the accurate accounting of reduced CO_2 emissions is difficult to be performed. Unlike geological storage, this technology captures CO_2, utilizes it as a product, and may be used in many applications. Therefore, overall quantity of CO_2 emissions reduction will be varied depending on the applications of carbonated products. The following points need to be preliminarily considered:

1. Utilization way of carbonated products should be measured in terms of the stability (CO_2 re-release extent).
2. Scenarios of the raw materials and energy source should be carefully confirmed.
3. Value chain management is required with a comprehensive life cycle assessment.
4. Definition of reporting boundary should be appropriately addressed.

If the CO_2 emissions reduction cannot be accurately assessed, this technology will encounter obstacles in the installation of Carbon Border Adjustment Mechanism, which may lead to errors in the calculation of carbon tax by this technology. In order to correctly assess the amount CO_2 emissions reduction, accounting methodology for various industries and applications should be established. In this section, a sample of mineral carbonation technology in cement industry was described.[18]

A calculation method of mineral carbonation technology in the cement industry had been established.[18] In this concept, the captured CO_2 is utilized to produce high-purity $CaCO_3$. The $CaCO_3$, also called recarbonate, is reused back to cement kiln to replace limestone to produce clinker. Although CO_2 will be released again when the recarbonate is calcined in a cement kiln, the amount of total CO_2 emissions to atmosphere is reduced because CO_2 is recycled in this system. Through this process, the unavoidable process-related emissions can be reduced, and the natural exploitation of limestone can be prevented.

The reporting boundaries for GHG emissions calculations are shown in Figure 4.6. The reporting boundaries can be classified in three parts: direct emission in cement plant, indirect emissions by power generation and transportations. The detailed information of input and output materials and CO_2 streams is shown in Figure 4.7. In a direct emission boundary, CO_2 emissions consideration includes emissions from raw materials such as limestone and recarbonate (1), emissions from fossil fuels and alternative fuels for a cement kiln (2), and CO_2 mineralization as recarbonate (3), and re-release from recarbonate (4). In an indirect emission boundary, CO_2 from purchased power is consumed by the mineral carbonation process (5), and CO_2 from transportation of alkaline materials (6). Through the above considerations, direct emissions, indirect emissions from imported energy, and other emissions along the value chain can be correctly calculated.

Based on the case sample, the establishment of the calculations method is essential to develop mineral carbonation technology. The use of carbonated products should be developed while taking the CO_2 balance throughout the entire boundary

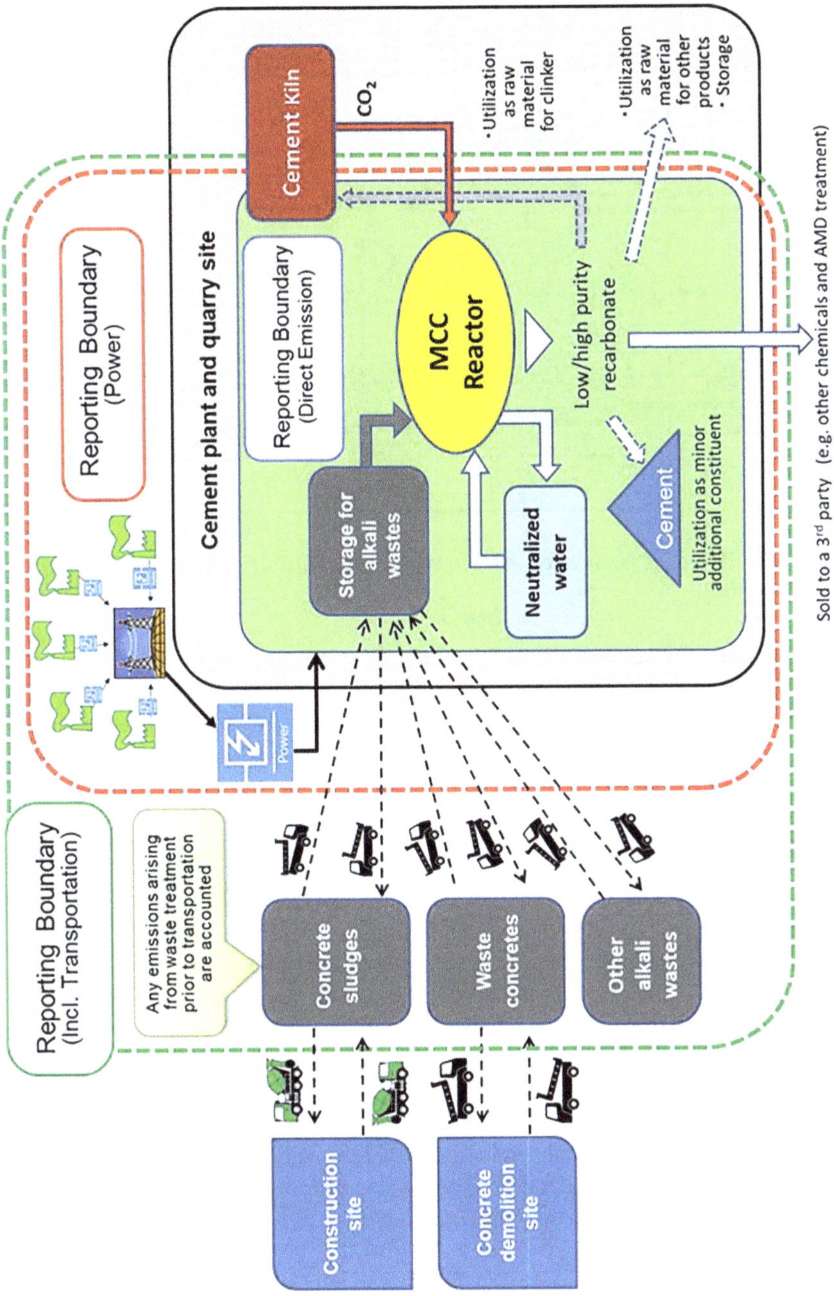

FIGURE 4.6 Reporting boundaries for GHG emissions calculations.[18]

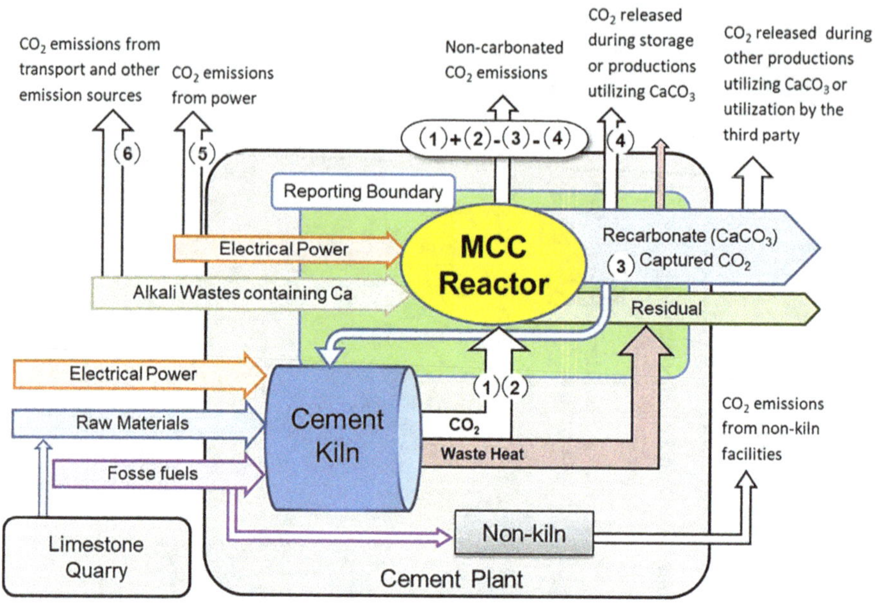

FIGURE 4.7 Reporting boundaries and cement plant site with all input and output materials and CO_2 streams.[18]

into account. The formulation of development strategies varies according to the type of industry. In addition, the energy structure and policies of individual countries need to be taken into account. In a nutshell, the calculation methods should be adjusted for different scenarios, conditions, and industries.

4.5 CONCLUSION

Mineral carbonation is a promising technology that can capture and utilize CO_2 as stable carbonates. The use of industrial waste as a raw material for mineral carbonation technology can both reduce CO_2 emissions and avoid the negative impact of the waste on the environment. Industries that generate alkaline waste, such as cement and steel industries, usually emit large volumes of CO_2 flue gas. Therefore, the idea of using industrial waste and unavoidable CO_2 gas via mineral carbonation of industrial waste is promising. According to different purposes and objectives, various carbonation processes can be developed. The calculation of CO_2 emissions reduction is complicated because CO_2 is not simply stored in geological reservoirs, but is captured as carbonated products and applied in further utilizations. In a nutshell, mineral carbonation of industrial waste is an important route for industries, in order to move toward carbon neutrality and a circular economy.

REFERENCES

1. Nataly Echevarria Huaman R, Xiu Jun T. 2014. Energy related CO_2 emissions and the progress on CCS projects: A review. *Renew Sust Energ Rev*. 31:368–385. doi:10.1016/j.rser.2013.12.002

2. International Energy Agency. 2020. *Energy Technology Perspectives 2020 - Special Report on Carbon Capture Utilisation and Storage*. Paris: OECD. doi:10.1787/208b66f4-en
3. Olabi AG, Wilberforce T, Elsaid K, Sayed ET, Maghrabie HM, Abdelkareem MA. 2022. Large scale application of carbon capture to process industries – A review. *J Clean Prod*. 362(May):132300. doi:10.1016/j.jclepro.2022.132300
4. Zhang Z, Pan S-Y, Li H, Cai J, Olabi AG, Anthony EJ, Manovic V. 2020. Recent advances in carbon dioxide utilization. *Renew Sust Energ Rev* 125(February):109799. doi:10.1016/j.rser.2020.109799
5. Ho H-J, Iizuka A, Shibata E. 2019. Carbon capture and utilization technology without carbon dioxide purification and pressurization: A review on its necessity and available technologies. *Ind Eng Chem Res*. 58(21):8941–8954. doi:10.1021/acs.iecr.9b01213
6. Ho H-J, Iizuka A, Shibata E, Tomita H, Takano K, Endo T. 2020. CO_2 utilization via direct aqueous carbonation of synthesized concrete fines under atmospheric pressure. *ACS Omega*. 5(26):15877–15890. doi:10.1021/acsomega.0c00985
7. Ho H-J, Iizuka A, Shibata E, Tomita H, Takano K, Endo T. 2021. Utilization of CO_2 in direct aqueous carbonation of concrete fines generated from aggregate recycling: Influences of the solid–liquid ratio and CO_2 concentration. *J Clean Prod*. 312(January):127832. doi:10.1016/j.jclepro.2021.127832
8. Ho H-J, Iizuka A, Shibata E. 2021. Chemical recycling and use of various types of concrete waste: A review. *J Clean Prod* 284:124785. doi:10.1016/j.jclepro.2020.124785
9. Ho H-J, Iizuka A, Kubo H. 2022. Direct aqueous carbonation of dephosphorization slag under mild conditions for CO_2 sequestration and utilization: Exploration of new dephosphorization slag utilization. *Environ Technol Innov*. 28:102905. doi:10.1016/j.eti.2022.102905
10. Ho H-J, Iizuka A, Shibata E. 2021. Utilization of low-calcium fly ash via direct aqueous carbonation with a low-energy input: Determination of carbonation reaction and evaluation of the potential for CO_2 sequestration and utilization. *J Environ Manag*. 288(February):112411. doi:10.1016/j.jenvman.2021.112411
11. Ho H, Iizuka A, Shibata E, Ojumu T. 2022. Circular indirect carbonation of coal fly ash for carbon dioxide capture and utilization. *J Environ Chem Eng*. 10(5):108269. doi:10.1016/j.jece.2022.108269
12. Chen J, Shen Y, Chen Z, Fu C, Li M, Mao T, Xu R, Lin X, Li X, Yan J. 2023. Accelerated carbonation of ball-milling modified MSWI fly ash: Migration and stabilization of heavy metals. *J Environ Chem Eng*. 11(2):109396. doi:10.1016/j.jece.2023.109396
13. Chen TL, Chen YH, Dai MY, Chiang PC. 2021. Stabilization-solidification-utilization of MSWI fly ash coupling CO_2 mineralization using a high-gravity rotating packed bed. *Waste Manag*. 121:412–421. doi:10.1016/j.wasman.2020.12.031
14. Abdul F, Iizuka A, Ho H-J, Adachi K, Shibata E. 2023. Potential of major by-products from non-ferrous metal industries for CO_2 emission reduction by mineral carbonation: A review. *Environ Sci Pollut Res*. (0123456789). doi:10.1007/s11356-023-27898-y
15. Puthiya Veetil SK, Rebane K, Yörük CR, Lopp M, Trikkel A, Hitch M. 2021. Aqueous mineral carbonation of oil shale mine waste (limestone): A feasibility study to develop a CO_2 capture sorbent. *Energy*. 221:119895. doi:10.1016/j.energy.2021.119895
16. Renforth P. 2019. The negative emission potential of alkaline materials. *Nat Commun*. 10(1):1401. doi:10.1038/s41467-019-09475-5
17. Ho H-J, Iizuka A, Lee C-H, Chen W-S. 2023. Mineral carbonation using alkaline waste and byproducts to reduce CO_2 emissions in Taiwan. *Environ Chem Lett*. 21(2):865–884. doi:10.1007/s10311-022-01518-6
18. Izumi Y, Iizuka A, Ho H-J. 2021. Calculation of greenhouse gas emissions for a carbon recycling system using mineral carbon capture and utilization technology in the cement industry. *J Clean Prod*. 312(December 2020):127618. doi:10.1016/j.jclepro.2021.127618

19. Rushendra Revathy TD, Palanivelu K, Ramachandran A. 2017. Sequestration of carbon dioxide by red mud through direct mineral carbonation at room temperature. *Int J Glob Warm.* 11(1):23. doi:10.1504/IJGW.2017.080988
20. Tian S, Jiang J, Chen X, Yan F, Li K. 2013. Direct gas-solid carbonation kinetics of steel slag and the contribution to insitu sequestration of flue gas CO_2 in steel-making plants. *ChemSusChem.* 6(12):2348–2355. doi:10.1002/cssc.201300436
21. Zajac M, Skibsted J, Bullerjahn F, Skocek J. 2022. Semi-dry carbonation of recycled concrete paste. *J CO_2 Util.* 63(June):102111. doi:10.1016/j.jcou.2022.102111
22. Zajac M, Maruyama I, Iizuka A, Skibsted J. 2023. Enforced carbonation of cementitious materials. *Cem Concr Res.* 174(August):107285. doi:10.1016/j.cemconres.2023.107285
23. Ho H, Iizuka A. 2023. Mineral carbonation using seawater for CO_2 sequestration and utilization: A review. *Sep Purif Technol.* 307:122855. doi:10.1016/j.seppur.2022.122855
24. Azdarpour A, Asadullah M, Mohammadian E, Hamidi H, Junin R, Karaei MA. 2015. A review on carbon dioxide mineral carbonation through pH-swing process. *Chem Eng J.* 279:615–630. doi:10.1016/j.cej.2015.05.064
25. Sanna A, Dri M, Maroto-Valer M. 2013. Carbon dioxide capture and storage by pH swing aqueous mineralisation using a mixture of ammonium salts and antigorite source. *Fuel.* 114:153–161. doi:10.1016/j.fuel.2012.08.014
26. Iizuka A, Sakai Y, Yamasaki A, Honma M, Hayakawa Y, Yanagisawa Y. 2012. Bench-scale operation of a concrete sludge recycling plant. *Ind Eng Chem Res.* 51(17):6099–6104. doi:10.1021/ie300620u
27. Iizuka A, Sasaki T, Honma M, Yoshida H, Hayakawa Y, Yanagisawa Y, Yamasaki A. 2017. Pilot-scale operation of a concrete sludge recycling plant and simultaneous production of calcium carbonate. *Chem Eng Commun.* 204(1):79–85. doi:10.1080/00986445.2016.1235564
28. Iizuka A, Ho H, Sasaki T, Yoshida H, Hayakawa Y, Yamasaki A. 2022. Comparative study of acid mine drainage neutralization by calcium hydroxide and concrete sludge-derived material. *Miner Eng.* 188(May):107819. doi:10.1016/j.mineng.2022.107819
29. Iizuka A, Fujii M, Yamasaki A, Yanagisawa Y. 2004. Development of a new CO_2 sequestration process utilizing the carbonation of waste cement. *Ind Eng Chem Res.* 43(24):7880–7887. doi:10.1021/ie0496176
30. Shuto D, Nagasawa H, Iizuka A, Yamasaki A. 2014. A CO_2 fixation process with waste cement powder via regeneration of alkali and acid by electrodialysis. *RSC Adv.* 4(38):19778–19788. doi:10.1039/C4RA00130C
31. Shuto D, Igarashi K, Nagasawa H, Iizuka A, Inoue M, Noguchi M, Yamasaki A. 2015. CO_2 fixation process with waste cement powder via regeneration of alkali and acid by electrodialysis: effect of operation conditions. *Ind Eng Chem Res.* 54(25):6569–6577. doi:10.1021/acs.iecr.5b00717
32. Taiwan Environmental Protection Administration. 2021. National Greenhouse Gas Inventory Report 2021.
33. Pan S, Chen Y, Fan L, Kim H, Gao X, Ling T, Chiang P, Pei S, Gu G. 2020. CO_2 mineralization and utilization by alkaline solid wastes for potential carbon reduction. *Nat Sustain.* 3(5):399–405. doi:10.1038/s41893-020-0486-9
34. Woodall CM, McQueen N, Pilorgé H, Wilcox J. 2019. Utilization of mineral carbonation products: Current state and potential. *Greenh Gases Sci Technol.* 9(6):1096–1113. doi:10.1002/ghg.1940

5 Photo(electro)chemical Systems for Upcycling of Carbon-Containing Waste

Fitri Nur Indah Sari, Yi-Hsuan Lai, and Chia-Yu Lin
National Cheng Kung University, Tainan City, Taiwan

5.1 INTRODUCTION

Rapid global population growth and industrialization over the past years have led to excessive resource extraction (e.g., fossil fuel), industrial goods production, and waste disposal. These human activities significantly increase the global energy demand and necessitate intensive use of fossil fuels, which results in an unprecedented increase in atmospheric CO_2 concentration and, in turn, triggers a series of catastrophic impacts, such as global warming and extreme weather. For instance, energy supply and its usage contribute to about 55% of anthropogenic greenhouse gas (GHG) emissions, whereas agriculture, industrial goods production, and waste disposal are responsible for the remaining 45% of anthropogenic GHG emissions (1). It is important to note that around two gigatons of municipal solid waste are generated globally by consumers each year, and most of them (~70%) accumulate in landfills or even escape into our environment every year, which also poses threats to public health, the environment, and the aquatic ecosystem (2). For example, the annual production of plastics increased by approximately 390.7 million tons in 2021 and is forecast to reach 1100 million tons by 2050 (3). However, only 6–26% of plastic wastes are recycled, while the remaining end up in landfills (4). These plastic wastes also contribute to environmental pollution in various places, including soils and oceans (5, 6). Several conventional waste management strategies have been applied for the treatment of these solid wastes, including gasification (7), pyrolysis (8), and fermentation (9). Nevertheless, these technologies include certain drawbacks, including the requirement for elevated temperatures and further post-treatment procedures to purify the resulting gas product effectively. Additionally, the fermentation approach is not deemed appropriate for the treatment of plastic waste.

A complete shift from fossil fuels to renewable energy sources (e.g., solar energy, wind power, hydropower) is expected to tackle 55% of GHG emissions, but renewable energies suffer intermittency and inability to produce heat or fuel, which currently causes their low contribution (<10%) to global energy production. Hydrogen is a clean energy carrier with high specific energy density, which makes itself a

DOI: 10.1201/9781003463009-5

promising vector bridging this gap and for the various applications ranging from carbon-free fuel and heating to the industrial synthesis of commodity chemicals. Hydrogen generation from photoelectrocatalytic (PEC) water splitting serves as a sustainable and carbon-free synthetic alternative to the conventional steam reforming of fossil fuels. PEC water splitting consists of two half-reactions, including hydrogen evolution generation (HER) via proton reduction (Eq. 5.1) and oxygen evolution reaction (OER) via water oxidation (Eq. 5.2). In these two half-reactions, OER involves a complicated four proton-coupled electron transfer processes and suffers sluggish kinetics. Consequently, it places a kinetic bottleneck in the overall water splitting process, which prevents substantial hydrogen generation. Replacement of OER with oxidative reforming of solid waste can be an effective strategy for efficient hydrogen generation, as the oxidative reforming of the solid waste is thermodynamically more favourable than OER (Figure 5.1). For example, the overall water splitting is a thermodynamically uphill process, with a Gibbs free energy change ($\Delta G°$) of $+237$ kJ mol^{-1}, whereas overall reforming of ethylene glycol (EG, monomeric unit of polyethylene terephthalate) or glucose (monomeric unit of carbohydrates) only has $\Delta G°$ of 9.2 kJ mol^{-1} (10) or -84.7 kJ mol^{-1} (11). In other words, PEC hydrogen generation coupled with the reforming of solid waste can further reduce the overall energy expense and enhance hydrogen generation compared to PEC water splitting. Additional benefits from PEC reforming of solid waste include mitigating the

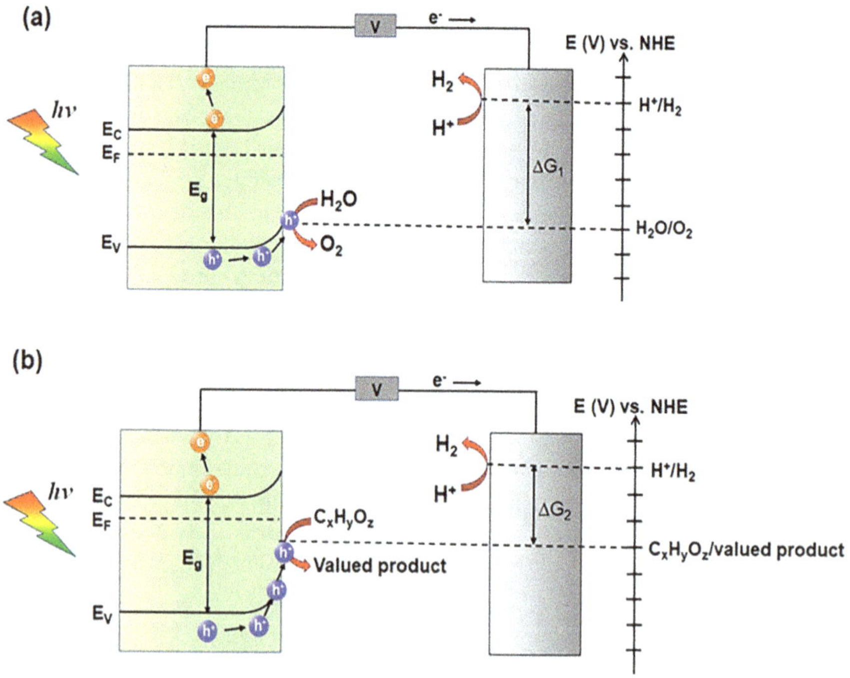

FIGURE 5.1 Schematic illustration of (a) a PEC waster splitting system and (b) a PEC system for the reforming of organic waste. E_C, E_V, and E_F are the conduction band edge, valence band edge, and Fermi-level of the semiconductor, respectively.

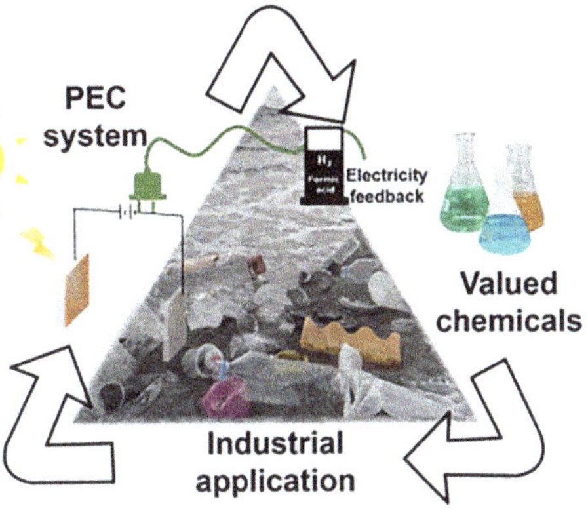

FIGURE 5.2 A renewable solar energy-driven photoelectrocatalytic (PEC) system to recycle and convert plastic waste into valued chemicals.

environmental pollution posed by plastics waste and upcycling plastic waste into valued chemicals, which enables a circular economy (Figure 5.2). For example, polyethylene terephthalate (PET) waste can be upcycled into formic acid and clean hydrogen gas via photocatalytic and PEC reforming (12–14). In addition, the upcycling of biomass or biomass-derived oxygenates into valued chemicals (e.g., formic acid, 2,5-furandicarboxylic acid) and hydrogen fuel by PEC reforming has also been demonstrated (15–18). Altogether, PEC reforming represents a potentially effective approach to address the above-mentioned challenges posed by increasing energy demand and environmental pollution from uncontrollable waste disposal.

$$2H^+ + 2e^- \rightarrow H_2 \qquad E^0 = 0.0\ V\ (vs.\ NHE) \tag{5.1}$$

$$2H_2O \rightarrow 4H^+ + 4e^- + O_2 \qquad E^0 = 1.23\ V\ (vs.\ NHE) \tag{5.2}$$

$$C_6H_6O_3 \rightarrow C_6H_4O_5 + 6H^+ + 6e^- \qquad E^0 = 0.11\ V\ (vs.\ NHE) \tag{5.3}$$

In this chapter, we will examine the recent progress and remaining challenges in developing the PEC reforming systems for upcycling biomass and plastic waste into value-added chemicals.

5.2 PHOTOELECTROCATALYTIC (PEC) REFORMING

5.2.1 Working Principle of PEC Reforming

PEC waste reforming couples HER half-reaction (Eq. 5.1) with oxidative reforming of organic wastes (e.g., oxidation of 5-hydroxymethylfurfural; Eq. 5.3) with sunlight as the sole energy input. Figure 5.3 shows a typical PEC reforming device. Upon light illumination, electron-hole pairs (e^-/h^+) are created inside the light-harvesting materials on the photoanode, and after charge separation, the minority charge carriers, *i.e.*, holes, are driven by the built-in electric field at the

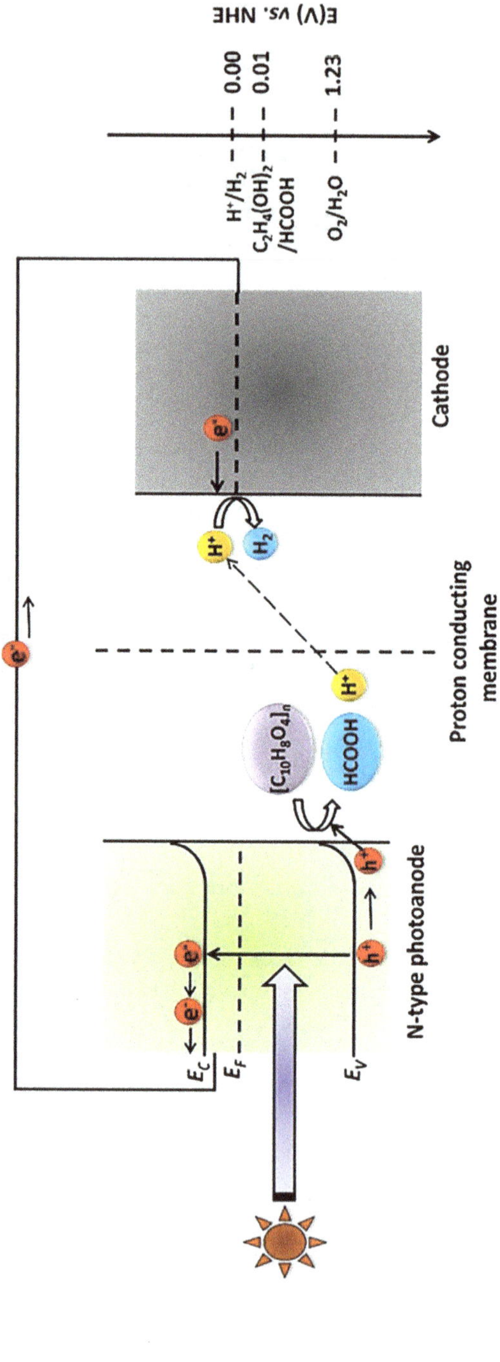

FIGURE 5.3 Schematic illustration of PEC system for the reforming of polyethylene terephthalate under solar light illumination. E_C, E_F, and E_V are conduction band edge, Fermi-level, and valence band edge of the n-type photoanode, respectively.

photoanode|electrolyte junction into the electrolyte solution to drive the oxidation of organic waste to produce the desired product (Figure 5.3), and (ii) the majority charge carriers, *i.e.*, electrons, transport through the bulk material and external circuit to the cathode to drive HER at the counter electrode. In other words, the overall PEC reforming process can be divided into three main steps, including light absorption, charge separation and transport of charge carriers, and interfacial catalysis. The external quantum efficiency (EQE) describing the whole process is shown in Eq. (5.4) (19, 20):

$$\mathrm{EQE}(\lambda) = \eta_{e^-/h^+} \times \eta_{trans} \times \eta_{\mathrm{int}} \tag{5.4}$$

where η_{e^-/h^+} is the fraction of electron-hole pairs generated per incident photon flux, η_{trans} is charge transport efficiency, and η_{int} is the interfacial charge transfer efficiency.

To drive the overall reforming process solely by solar energy, the band structure and intrinsic properties of the light-harvesting material should meet several energetics criteria (Figure 5.4). To begin with, the energy level of the bottom edge of the conduction band of the light-harvesting material should be more negative than the reduction potential of HER (0.0 V vs. RHE) so that photogenerated electrons would have the sufficient reductive equivalent for HER. In addition, the energy level of the upper edge of the valence band of the light-harvesting material should be more positive than the reduction potential of organic waste oxidation reaction (e.g., 0.01 V vs. RHE for EG, a monomeric unit of PET (10)) so that photogenerated holes would have sufficient oxidative equivalent to oxidize organic waste. This means that only SC1 and SC2 semiconducting materials in Figure 5.4 can potentially drive the PEC reforming of organic waste into hydrogen and valued chemicals with solar energy as the sole energy input. Furthermore, the bandgap (i.e., the distance between the bottom edge of the conduction band and the upper edge of the valence band) should be sufficiently small to harvest a larger portion of solar spectrum and thus maximize η_{e^-/h^+}. This means that SC2 would have higher η_{e^-/h^+} than SC1 in Figure 5.4. Finally, the recombination of the photogenerated charge carriers both in the bulk (bulk recombination; path (iii) in Figure 5.5) and surface (surface recombination; path (iv) in Figure 5.5) of the light-harvesting material should be minimal to ensure high η_{trans} and η_{int}.

Unfortunately, most of the semiconducting light-harvesting materials suffer server charge recombination. The bulk recombination is mainly attributed to the intrinsic low conductivity of the semiconducting light-harvesting materials. That is, the transport kinetics of the majority charge carrier is slow, while the surface recombination is highly related to the sluggish interfacial transfer of the minority charge carriers (19, 21–25). The charge recombination becomes severe, especially when the effective charge diffusion length is not commensurate with the film thickness required to fully harvest photons with energy close to the bandgap of the light-harvesting material (i.e., the light-absorption length). For example, the incommensurate light-absorption length with electron diffusion length, where the light absorption length near the bandgap of Cu_2O (~10 μm) (26, 27) is much higher than that of diffusion length of the electron inside Cu_2O (~10–100 nm) (28), for p-type Cu_2O has also been considered as one of the main reasons why the maximal photocurrent density

photoanode towards the production of formic acid with high Faradaic efficiency (~95%) from PEC methanol reforming at near-neutral pH (Figure 5.7) (32). Cha et al. (15) reported that the bare $BiVO_4$ photoanode is poorly active for water oxidation but is highly catalytic for the PEC oxidation of 2,2,6,6-tetramethylpiperidine-1-oxyl (TEMPO), which is a mediator and catalyst in catalyzing the oxidative conversion of 5-hydroxymethylfurfural (HMF) into furandicarboxylic acid (FDCA), a key platform molecule for the synthesis of important polymeric materials. In other words, the inclusion of TEMPO in the HMF-containing electrolyte solution of the $BiVO_4$-based PEC system can fully suppress water oxidation, and the photogenerated holes can be effectively directed to the oxidation of TEMPO into $TEMPO^+$ (Eq. 5.5), which in turn selectively oxidizes HMF into FDCA (Eq. 5.6). The generation of FDCA with 100% Faradaic efficiency and a near-quantitative yield was obtained with this strategy.

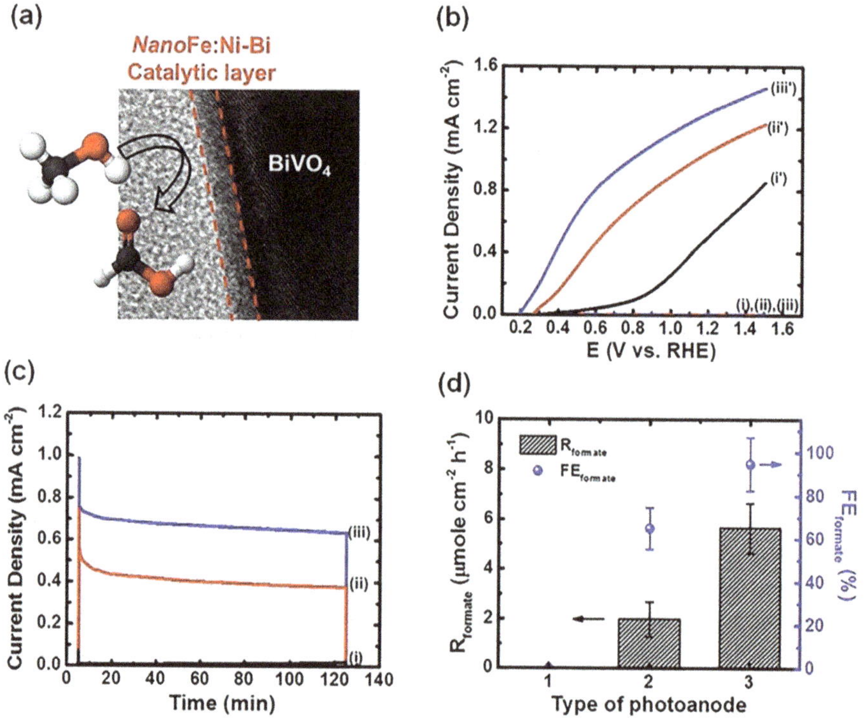

FIGURE 5.7 (a) The surface-modification of the $BiVO_4$ photoanode with nanoFe:NiBi. (b) LSV curves (scan rate of 10 mV s^{-1}) of (i, i′) the pristineBiVO$_4$, (ii, ii′) $BiVO_4$|*nano*Ni-Bi, and (iii, iii′) $BiVO_4$|*nano*Fe:Ni-Bi$_{(r=0.11)}$ photoanodes in borate buffer (0.1 M, pH 9.4) containing methanol (0.1 M) in the dark (i′, ii′, iii′) and under light illumination (i, ii, iii). (c) Photocurrent transients recorded at an applied potential of 0.55 V vs. RHE, of (i) the pristine $BiVO_4$, (ii) $BiVO_4$|*nano*Ni-Bi, and (iii) $BiVO_4$|*nano*Fe:Ni-Bi$_{(r=0.11)}$ photoanodes in borate buffer (0.1 M, pH 9.4) containing methanol (0.1 M) under light illumination (100 mW cm^{-2}). (d) $R_{formate}$ and $FE_{formate}$ from 2-h PEC methanol reforming at the pristine $BiVO_4$ (#1), $BiVO_4$|*nano*Ni-Bi (#2), and $BiVO_4$|*nano*Fe:Ni-Bi$_{(r=0.11)}$ (#3) photoanodes.

(Reproduced from Ref. (32) with permission from Elsevier B.V.)

$$\text{TEMPO} + h^+ \rightarrow \text{TEMPO}^+ \tag{5.5}$$

$$\text{TEMPO}^+ + \text{HMF} \rightarrow \text{FDCA} + \text{TEMPO} \tag{5.6}$$

5.2.2 State-of-the-Art PEC Waste Reforming

As discussed previously, the success in the establishment of the efficient PEC reforming system relies on the development of two key components, including (i) light-harvesting material, which can capture a large portion of solar energy and generates energetic charge carriers with minimal charge recombination, and (ii) the robust electrocatalyst, which can efficiently expedite the kinetics of reactions of interests and generate specific products with high selectivity. Chuang et al. investigated the electrocatalytic activity of monoclinic CuO (*m*-CuO) towards the electrochemical oxidation of glucose, cellulose, and rice straw, and they found that the oxidation of these biomass waste or biomass-derived oxygenate involves a C–C bond cleavage, mediated by the generation of OH radical, and generates formate as the main product (33); formate production with Faradaic efficiencies ($FE_{formate}$) of 94.1 ± 1.5% and 41.4 ± 9.7% was obtained from the electrochemical oxidation of glucose and rice straw, respectively. Encouraged by the high activity of *m*-CuO, they also integrated *m*-CuO onto the hematite nanorods photoanode (nanoFe_2O_3) to establish an efficient PEC system for glucose reforming (Figure 5.8). They found that the bare nanoFe_2O_3 photoanode suffered low Faradaic efficiency due to the competition of OER during the PEC glucose reforming. In contrast, the surface modification of the nanoFe_2O_3 with *m*-CuO can direct the photogenerated holes to the oxidation process of glucose and generate formate with high $FE_{formate}$ (97.3 ± 2.8%).

Recently, $CuBi_2O_4$, a p-type semiconductor, has been explored as an efficient electrocatalyst in catalyzing the electrochemical reforming of glucose to generate formate with high selectivity (34, 35). Very recently, our group investigated the PEC performance of the $CuBi_2O_4$-modified TiO_2 photoanode ($TiO_2|CuBi_2O_4$) towards the reforming of cellulose. The fabrication of the $TiO_2|CuBi_2O_{42}$ photoanode consisted of three steps, including (i) the preparation of TiO_2 paste, by dispersing 0.5 g Degussa P25 TiO_2 nanopowder in 2.6 mL de-ionised water containing ~4 vol% Triton™ X-100 under vigorous agitation, (ii) the fabrication of TiO_2 photoanode by casting TiO_2 paste onto the fluorine-doped tin oxide coated glass substrate (FTO) by doctor-blade method and subsequent thermal annealing at 500°C for 1 h, and (iii) drop-coating an ethanolic precursor solution (10 μL cm^{-2}) containing 2 mM copper acetate and 4 mM bismuth nitrate onto the TiO_2 photoanode, followed by thermal annealing at 500°C for 3 h. To adjust the amount of $CuBi_2O_4$, the step of the drop-coating of the $CuBi_2O_4$ precursor solution was repeated for different times (*n*). The unpublished results (Figure 5.9a) reveal that the $TiO_2|CuBi_2O_4$ photoanodes exhibited lower photocurrent responses than the bare TiO_2 photoanode in the blank 0.1 M NaOH solution, which indicates that the surface modification of TiO_2 with $CuBi_2O_4$ effectively switched off OER. In addition, as revealed in Figure 5.9b, the $TiO_2|CuBi_2O_4$ photoanodes exhibited significantly enhanced photocurrent responses in α-cellulose (68 mg L^{-1})-containing NaOH solution (0.1 M), which suggests that the surface-modification of TiO_2 with $CuBi_2O_4$ offers active sites to facilitate the PEC cellulose reforming.

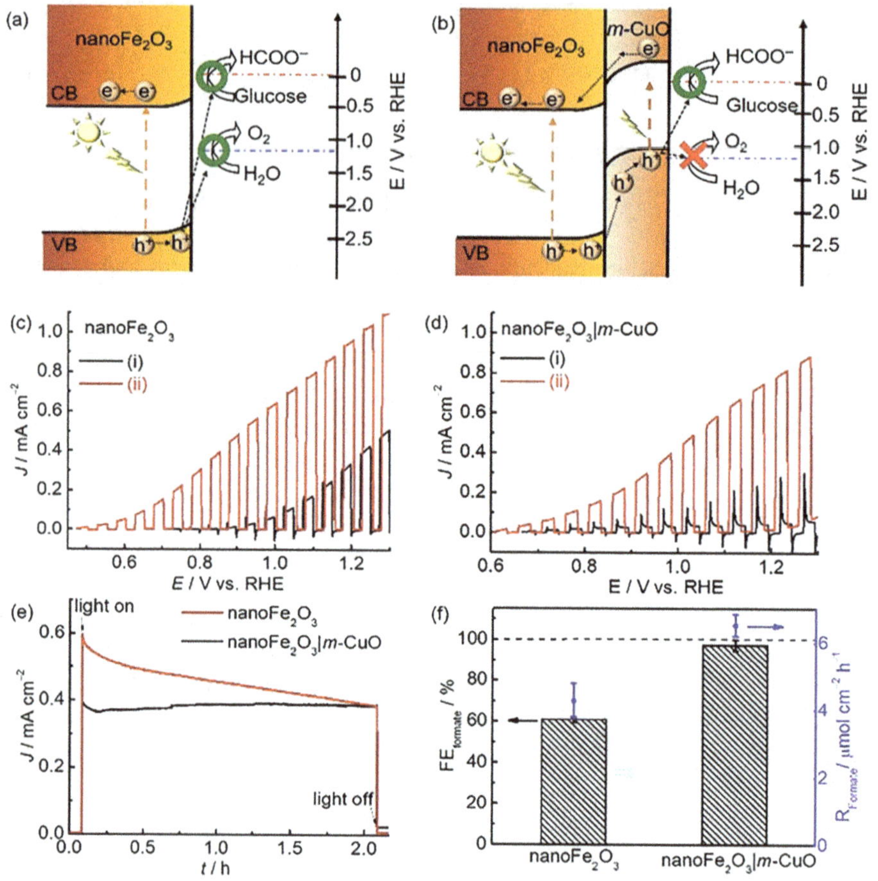

FIGURE 5.8 Schematic illustration of the electronic band structures of (a) nanoFe_2O_3 and (b) nanoFe_2O_3|m-CuO. LSV curves with a scan rate of 5 mV s^{-1} of (c) nanoFe_2O_3 and (d) nanoFe_2O_3|m-CuO recorded in (i) 0.1 M NaOH solution and (ii) 0.1 M NaOH solution containing 20 mM glucose under chopped solar light irradiation. (e) Chronopotentiometry measurements and (f) $FE_{formate}$ and $R_{formate}$ of nanoFe_2O_3 and nanoFe_2O_3|m-CuO at 1.0 V vs. RHE in 0.1 M NaOH solution containing 20 mM glucose under solar light irradiation.

(Reproduced from Ref. (33) with permission from the Royal Society of Chemistry.)

However, the TiO_2|$CuBi_2O_4$ photoanodes exhibited decreasing photocurrent responses with increasing loading amount of $CuBi_2O_4$, which could be attributed to the deteriorated kinetics of charge transport in bulk due to the unfavorable band alignment. The results (Figures 5.9(c, d)) of 2-h prolonged photoelectrolysis experiments, at 0.5 V vs. RHE in 0.1 M NaOH solution containing 68 mg L^{-1} α-cellulose, confirm that the enhanced photocurrent responses were resulted from PEC cellulose reforming. It can be found that the bare TiO_2 photoanode (i.e., $n = 0$) exhibited the highest photocurrent response, but lowest $FE_{formate}$ (25.19 ± 1.67%) and $R_{formate}$ (0.90 ± 0.06 μmol cm^{-2} h^{-1}). In contrast, the modification of TiO_2 photoanode with a small amount of $CuBi_2O_4$,

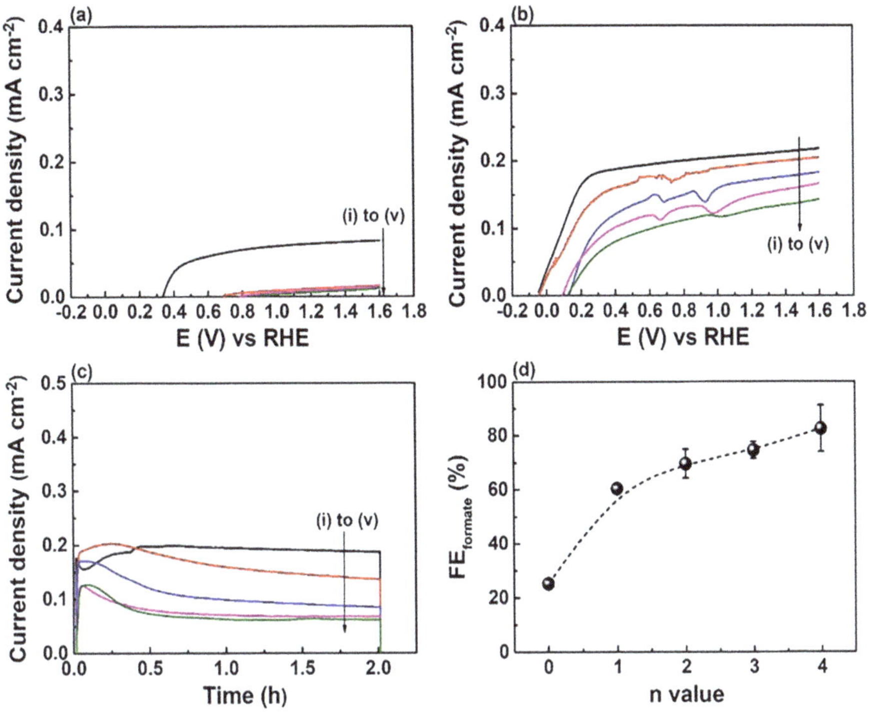

FIGURE 5.9 PEC performance of the $TiO_2|CuBi_2O_4$ electrode. (a, b) LSV curves, recorded at a scan rate of 10 mV s^{-1}, of the $TiO_2|CuBi_2O_4$ electrodes prepared in 0.1 M NaOH solution (a) and in cellulose-containing 0.1 M NaOH solution (b). (c) Photocurrent transients of the $TiO_2|CuBi_2O_4$ electrodes recorded in the 2-h controlled-potential photoelectrolysis experiments at 0.5 V vs. RHE in cellulose-containing 0.1 M NaOH solution. Sample id in (a–c): (i): n = 0, (ii): n = 1, (iii): n = 2, (iv): n = 3, (v): n = 4. (d) $FE_{formate}$ obtained from 2-h photoelectrolysis experiments in (c).

i.e., $TiO_2|CuBi_2O_4$ prepared with $n = 1$, significantly improved $FE_{formate}$ (60.38 ± 1.53%) and $R_{formate}$ (1.71 ± 0.15 μmol cm^{-2} h^{-1}). Further increase in the loading amount of $CuBi_2O_4$ (i.e., increasing n) further improved $CE_{formate}$ up to 82.68 ± 8.51%, but significantly decreased the overall photocurrent response, resulting in the decrease in $R_{formate}$.

In the case of PEC plastics reforming, Lin et al. recently established a PEC system, consisting of a cathode made with the nanocomposite of carbon nanotubes with nickel phosphide nanoparticles (CNT/nanoNiP) and a nickel phosphide nanoparticles-modified TiO_2 nanorods photoanode (nanoTiO_2/nanoNiP), for the reforming of PET waste into formate and hydrogen (Figure 5.10) (14). In their work, they applied CV-pretreatment to render the surfaces of nanoNiP and nickel nanoparticles with an ultrathin layer of nickel oxyhydroxide (NiOOH). In addition, $P^{\delta-}$ in nanoNiP was found to play a vital role in suppressing further transformation of β-NiOOH into γ-NiOOH, resulting in the higher content of β-NiOOH and electrocatalytic activity of the CV-pretreated nanoNiP as β-NiOOH is the active species responsible for the

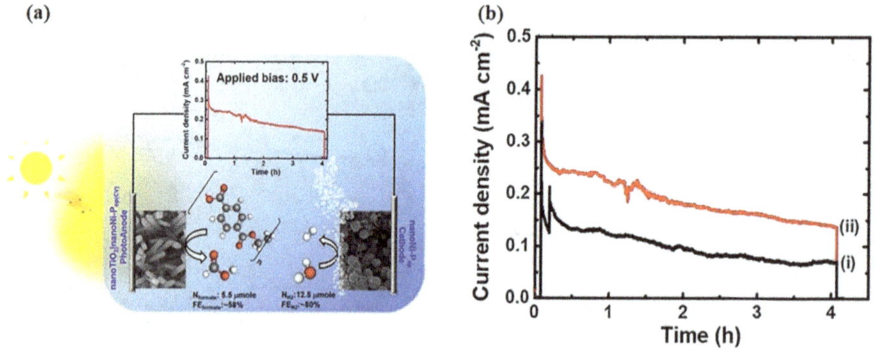

FIGURE 5.10 (a) Schematic illustration of the nanoTiO$_2$/nanoNi-P||CNT/nanoNi-P PEC system. (b) Photocurrent transients recorded during the PEC PET reforming at an external bias of 0.5 V using (i) the nanoTiO$_2$/nanoNiP||CNT/nanoNiP and (ii) the nanoTiO$_2$||Pt PEC systems.

(Reproduced from Ref. (14) with permission from Elsevier B. V.)

electrochemical and PEC forming of EG and PET. The CV-pretreated nanoNiP exhibited high selectivity towards the generation of formate (FE$_{formate}$ = ~100%) from the electrochemical reforming of PET. Furthermore, the pristine nanoNiP was found to be highly active towards HER, requiring a small overpotential of −180 mV to drive HER at −100 mA cm^{-2}. Encouraged by these promising electrocatalytic activities, nanoNiP was further integrated onto the TiO$_2$ nanorods photoanode for the PEC reforming of PET. In the 4-h photoelectrolysis at a bias of 0.5 V (Figure 5.10(b)), the nanoTiO$_2$/nanoNiP||CNT/nanoNiP PEC system generated 12.5 ± 1.1 μmol H$_2$ (FE$_{hydrogen}$ of 76.8 ± 7.8 %) and 6.4 ± 1.4 μmol formate (FE$_{formate}$ of 57.1 ± 1.7 %). In contrast, the nanoTiO$_2$||Pt PEC system only generated 7.1 ± 2.4 μmol H$_2$(FE$_{hydrogen}$: 65.7 ± 11.7 %) and 1.0 ± 0.1 μmol formate (FE$_{formate}$: 14.7 ± 2.5 %). The significantly higher FE$_{formate}$ of nanoTiO$_2$/nanoNiP photoanode than that of nanoTiO$_2$ photoanode highlights the critical role of nanoNiP.

Reisner and his coworkers have recently successfully established a perovskite-based PEC reforming system for upcycling biomass and plastic wastes into value-added chemicals and hydrogen with solar energy as the sole energy input (36). This reforming PEC system consisted of a Cu$_{30}$Pd$_{70}$ alloy micro-flowers anode for waste oxidation and a Pt-modified lead-halide perovskite photocathode for HER (Figures 5.11(a, b)). Without any external bias, the developed PEC device exhibited photocurrents of 4–9 mA cm^{-2} for the reforming of plastic, biomass, and glycerol with 60–90% product selectivity in either a two-compartment or integrated "artificial leaf" configuration (Figures 5.11(c–e)). In addition, this bias-free PEC device exhibited a 2–4 order of magnitude enhancement in the product rates over their photocatalytic counterparts, such as carbon nitride, carbon dots, and cadmium sulfide. This study shows the benefits of PEC devices over the existing photocatalytic reforming systems, making a big step towards widespread waste utilization.

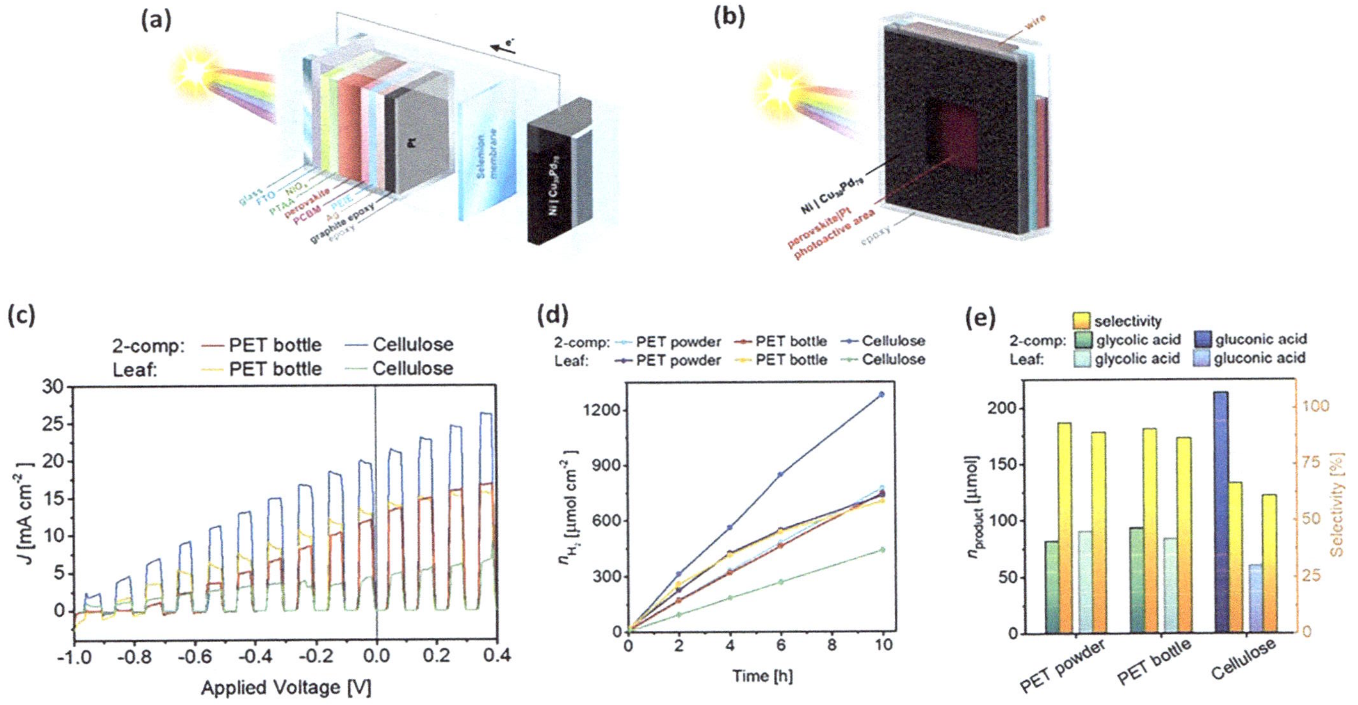

FIGURE 5.11 Schematic depictions of the $Cu_{30}Pd_{70}$|perovskite|Pt system in two-compartment (a), and standalone "artificial leaf" configurations (b). (c) Forward CV scans recorded at a scan rate of 10 mV s^{-1} under chopped simulated solar light irradiation (AM 1.5G, 100 mW cm^{-2}). (d) The amount of H_2 produced at different time intervals during the 10 h CA experiment at zero applied voltage. (e) The amount of oxidation product formed from the respective substrates and the corresponding selectivity after 10 h of bias-free measurement.

(Reproduced from Ref. (36) with permission from John Wiley and Sons.)

5.3 CONCLUSION AND OUTLOOK

PEC reforming has been proven as a green and sustainable technology that can simultaneously mitigate waste and upcycle waste into valuable chemicals and clean fuels, with solar energy as the sole energy input. Nevertheless, there remains fundamental challenges for the establishment of high-performance of the photoelectrochemical devices for upcycling the diverse wastes, especially for emerging pollutants, engineering plastics (e.g., polyethylene, polypropylene), CO_2, and even poly-fluoroalkyl substances. For example, the solubility of these polymeric materials is limited, and additional pretreatment steps should be designed for the specific wastes. In addition, the solar-to-chemical efficiency of the PEC device requires further improvement, including (i) the development of light-harvesting materials with high light-harvesting capability and minimal energy loss due to the transport kinetics of photogenerated charge carriers and (ii) the robust electrocatalyst, which can efficiently expedite the kinetics of reactions of interests, generate specific products with high selectivity, and be prepared and integrated with photoelectrodes by simple and scalable methods. With continued development and integration with other renewable technologies, PEC reforming is expected to play an important role in enabling the circular and sustainable flow of materials and energy and thus realizing a carbon-neutral future.

REFERENCES

1. MacArthur E Completing the picture - How the circular economy tackles climate change. Ellen MacArthur Found. 2019.
2. Uekert T, Pichler CM, Schubert T, Reisner E. Solar-driven reforming of solid waste for a sustainable future. *Nat Sustain*. 2021;4(5):383–91.
3. Plastics – The Facts 2022. https://plasticseurope.org/knowledge-hub/plastics-the-facts-2022
4. Alimi OS, Farner Budarz J, Hernandez LM, Tufenkji N. Microplastics and nanoplastics in aquatic environments: Aggregation, deposition, and enhanced contaminant transport. *Environ Sci Technol*. 2018;52(4):1704–24.
5. Thompson RC, Olsen Y, Mitchell RP, Davis A, Rowland SJ, John AWG, et al. Lost at sea: Where is all the plastic? *Science*. 2004;304(5672):838.
6. Huang J, Chen H, Zheng Y, Yang Y, Zhang Y, Gao B. Microplastic pollution in soils and groundwater: Characteristics, analytical methods and impacts. *Chem Eng J*. 2021; 425:131870.
7. Couto N, Silva V, Monteiro E, Rouboa A. Exergy analysis of Portuguese municipal solid waste treatment via steam gasification. *Energy Convers Manag*. 2017;134:235–46.
8. Fivga A, Dimitriou I. Pyrolysis of plastic waste for production of heavy fuel substitute: A techno-economic assessment. *Energy*. 2018;149:865–74.
9. Tian H, Li J, Yan M, Tong YW, Wang C-H, Wang X. Organic waste to biohydrogen: A critical review from technological development and environmental impact analysis perspective. *Appl Energy*. 2019;256:113961.
10. Akhundi A, Naseri A, Abdollahi N, Samadi M, Moshfegh A. Photocatalytic reforming of biomass-derived feedstock to hydrogen production. *Res Chem Intermed*. 2022;48(5):1793–811.
11. Uekert T, Kasap H, Reisner E. Photoreforming of nonrecyclable plastic waste over a carbon nitride/nickel phosphide catalyst. *J Am Chem Soc*. 2019;141(38):15201–10.

12. Li X, Wang J, Zhang T, Wang T, Zhao Y. Photoelectrochemical catalysis of waste polyethylene terephthalate plastic to coproduce formic acid and hydrogen. *ACS Sustain Chem Eng*. 2022;10(29):9546–52.
13. Han M, Zhu S, Xia C, Yang B. Photocatalytic upcycling of poly(ethylene terephthalate) plastic to high-value chemicals. *Appl Catal B Environ*. 2022;316:121662.
14. Lin C-Y, Huang S-C, Lin Y-G, Hsu L-C, Yi C-T. Electrosynthesized Ni-P nanospheres with high activity and selectivity towards photoelectrochemical plastics reforming. *Appl Catal B Environ*. 2021;296:120351.
15. Cha HG, Choi K-S. Combined biomass valorisation and hydrogen production in a photoelectrochemical cell. *Nat Chem*. 2015;7(4):328–33.
16. Kasap H, Achilleos DS, Huang A, Reisner E. Photoreforming of lignocellulose into H_2 using nanoengineered carbon nitride under benign conditions. *J Am Chem Soc*. 2018;140(37):11604–7.
17. Nguyen V-C, Ke N-J, Nam LD, Nguyen B-S, Xiao Y-K, Lee Y-L, et al. Photocatalytic reforming of sugar and glucose into H_2 over functionalised graphene dots. *J Mater Chem A*. 2019;7(14):8384–93.
18. Shi C, Kang F, Zhu Y, Teng M, Shi J, Qi H, et al. Photoreforming lignocellulosic biomass for hydrogen production: Optimised design of photocatalyst and photocatalytic system. *Chem Eng J*. 2023;452:138980.
19. Zhong DK, Choi S, Gamelin DR. Near-complete suppression of surface recombination in solar photoelectrolysis by "Co-Pi" catalyst-modified W:$BiVO_4$. *J Am Chem Soc*. 2011;133(45):18370–7.
20. Chen ZB, Jaramillo TF, Deutsch TG, Kleiman-Shwarsctein A, Forman AJ, Gaillard N, et al. Accelerating materials development for photoelectrochemical hydrogen production: Standards for methods, definitions, and reporting protocols. *J Mater Res*. 2010;25(1):3–16.
21. Jo WJ, Jang JW, Kong KJ, Kang HJ, Kim JY, Jun H, et al. Phosphate doping into monoclinic $BiVO_4$ for enhanced photoelectrochemical water oxidation activity. *Angew Chem Int Ed*. 2012;51(13):3147–51.
22. Luo WJ, Yang ZS, Li ZS, Zhang JY, Liu JG, Zhao ZY, et al. Solar hydrogen generation from seawater with a modified $BiVO_4$ photoanode. *Energy Environ Sci*. 2011;4(10):4046–51.
23. Parmar KPS, Kang HJ, Bist A, Dua P, Jang JS, Lee JS. Photocatalytic and photoelectrochemical water oxidation over metal-doped monoclinic $BiVO_4$ photoanodes. *ChemSusChem*. 2012;5(10):1926–34.
24. Pilli SK, Furtak TE, Brown LD, Deutsch TG, Turner JA, Herring AM. Cobalt-phosphate (Co-Pi) catalyst modified Mo-doped $BiVO_4$ photoelectrodes for solar water oxidation. *Energy Environ Sci*. 2011;4(12):5028–34.
25. Park HS, Kweon KE, Ye H, Paek E, Hwang GS, Bard AJ. Factors in the metal doping of $BiVO_4$ for improved photoelectrocatalytic activity as studied by scanning electrochemical microscopy and first-principles density-functional calculation. *J Phys Chem C*. 2011;115(36):17870–9.
26. Engel CJ, Polson TA, Spado JR, Bell JM, Fillinger A. Photoelectrochemistry of porous p-Cu_2O films. *J Electrochem Soc*. 2008;155(3):F37–F42.
27. Sculfort JL, Guyomard D, Herlem M. Photoelectrochemical characterisation of the para-Cu_2O-non aqueous-electrolyte junction. *Electrochim Acta*. 1984;29(4):459–65.
28. de Jongh PE, Vanmaekelbergh D, Kelly JJ. Photoelectrochemistry of electrodeposited Cu_2O. *J Electrochem Soc*. 2000;147(2):486–9.
29. Paracchino A, Laporte V, Sivula K, Graetzel M, Thimsen E. Highly active oxide photocathode for photoelectrochemical water reduction. *Nat Mater*. 2011;10(6):456–61.

30. Peng Q, Kalanyan B, Hoertz PG, Miller A, do Kim H, Hanson K, et al. Solution-processed, antimony-doped tin oxide colloid films enable high-performance TiO_2 photoanodes for water splitting. *Nano Lett.* 2013;13(4):1481–8.
31. Stefik M, Cornuz M, Mathews N, Hisatomi T, Mhaisalkar S, Gratzel M. Transparent, conducting Nb:SnO_2 for host-guest photoelectrochemistry. *Nano Lett.* 2012;12(10):5431–5.
32. Huang S-C, Cheng C-C, Lai Y-H, Lin C-Y. Sustainable and selective formic acid production from photoelectrochemical methanol reforming at near-neutral pH using nanoporous nickel-iron oxyhydroxide-borate as the electrocatalyst. *Chem Eng J.* 2020;395: 125176.
33. Chuang P-C, Lai Y-H. Selective production of formate over a CuO electrocatalyst by electrochemical and photoelectrochemical biomass valorisation. *Catal Sci Technol.* 2022;12(21):6375–83.
34. Wu C-H, Onno E, Lin C-Y. CuO nanoparticles decorated nano-dendrite-structured $CuBi_2O_4$ for highly sensitive and selective electrochemical detection of glucose. *Electrochim Acta.* 2017;229:129–40.
35. Lin C-Y, Lin S-Y, Tsai M-C, Wu C-H. Facile room-temperature growth of nanostructured $CuBi_2O_4$ for selective electrochemical reforming and photoelectrochemical hydrogen evolution reactions. *Sustain Energy Fuels.* 2020;4(2):625–32.
36. Bhattacharjee S, Andrei V, Pornrungroj C, Rahaman M, Pichler CM, Reisner E. Reforming of soluble biomass and plastic derived waste using a bias-free $Cu_{30}Pd_{70}$|Perovskite|Pt photoelectrochemical device. *Adv Funct Mater.* 2022;32(7):2109313.

6 Tuning Light-Matter Interaction with Photonic Architectures for CO_2 Reduction

Wen-Hui Cheng
National Cheng Kung University, Tainan City, Taiwan

6.1 INTRODUCTION

Carbon dioxide as one of the dominant house heating gases has been considered to be a serious environmental hazard, resulting in an overall increase of the earth's temperature and threatening the life of all creatures. The transformation from fossil fuel to renewable energy source is necessary. The sustainable development goals (SDGs) developed by the United Nations General Assembly (UNGA) address the associated issue of affordable and clean energy within SDG 7 and climate action within SDG 13. Solar energy has drawn more and more attention during the past decade, due to the unlimited energy source and low geometric restriction. However, the low-capacity factor for solar energy is a prominent concern that deems energy storage to be necessary. A photon generated carrier can be utilized for chemical reaction in the photo(electro)chemical system. In this chapter, we will focus on how the knowledge of optic design and photon management can play a role in enhancing light matter interaction, specifically for facilitating carbon dioxide reduction reaction (CO_2RR).

6.2 FUNDAMENTALS OF PHOTO-DRIVEN CO_2 REDUCTION

CO_2RR can not only transform waste into chemical fuel with a higher energy density than H_2, but also can become an important strategy to save the world from global warming. As compared to hydrogen evolution reaction (HER), there are many challenges involved in the intrinsic catalytic properties of catalysts and integration difficulties for CO_2RR. The complexity of the kinetic landscape and adjacent thermodynamic energy between different reaction products makes it significantly challenging to control the selectivity. The thermodynamic potential versus reverse hydrogen electrode (RHE) and the charges involved in these reactions are listed in Table 6.1.

For the basic photo-driven chemical system, the optical properties (carrier generation) and charge transfer process play a significant role in conducting the final chemical reaction. Essentially, the energy of generated carriers would need to be large to exceed the thermodynamic potential between cathode (reduction) and anode

DOI: 10.1201/9781003463009-6

TABLE 6.1
Charge and Potential Information for Different Reduction Reactions

Product	e^-	*E* vs RHE (V)
H_2	2	0
CO	2	−0.11
CH_4	8	0.17
C_2H_4	12	0.08
C_2H_6	14	0.14
HCOOH	2	−0.17
CH_3COOH	8	0.13
C_2H_5OH	12	0.08
C_3H_7OH	18	0.10

(oxidation) reaction. Two types of configurations are commonly considered, including wired PV-EC (photovoltaic and electrolyzer are separated but series connected) and integrated PEC (catalysts are stacked on the light absorber), as illustrated in Figure 6.1. In the wired PV-EC system, each component can be engineered independently, reducing the complexity of integration and improving the stability. However, with proper design, higher overall efficiency can be expected in the integrated PEC system. Therefore, minimized shadowing loss from metal catalysts to realize sufficient absorption for charge generation is necessary in the integrated system.

Compared with conventional photoelectrochemical devices composed of semiconductors, in which light absorbers and catalysts are usually decoupled, plasmonic

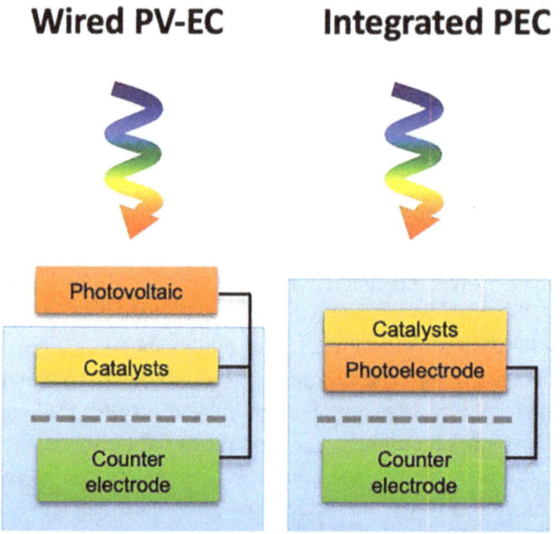

FIGURE 6.1 Illustration of two configurations in photo-driven chemical system.

photocatalysts offer unique size-dependent optical properties combined with catalytic surfaces that can influence chemical reactivity at the nanoscale, owing to the energetic hot carrier. Metallic nanostructures with surface plasmon resonance (SPR) that can harvest sunlight and generate non-equilibrium hot carriers capable of catalyzing chemical reactions offer an appealing platform for solar fuel generators. Especially interesting for CO_2RR, where the complex multi-step chemical pathway influences product selectivity, it has been proposed that plasmon excitation may reduce activation barriers, or change the population of adsorbed molecules.

The working principle of a plasmonic device for photocatalytic reaction is addressed below. The plasmonic material first adsorbs light upon illumination. Hot carriers are then generated. If the junction between metal and semiconductor presents to be Schottky, only hot carriers with enough energy can overcome the barrier. Such a barrier can prevent backflow of charges and recombination. Once the specific charges are extracted to the semiconductor for specific reaction (electrons for reduction and holes for oxidation), the alternative charges can then perform the other half-reaction accordingly.

6.3 DESIGN PRINCIPLE

To enable high solar-to-fuel efficiency, an optically transparent but active catalyst layer is needed. In previous work, a high solar-to-hydrogen PEC conversion efficiency of 19.3% is realized by integrating Rh nanoparticle catalysts onto photocathodes with minimal parasitic absorption and reflection losses in the visible range [1]. However, different strategies are applied to photo-driven CO_2R due to different material selection criteria. Figure 6.2 presents four concepts, including plasmonic tendency, toxicity, cost, and reduction product, that can be considered as the choice of catalysts in periodic table. Especially interesting for catalytic reaction, elements with more than half of the d-band states filled (>5 d-band electrons) are discussed.

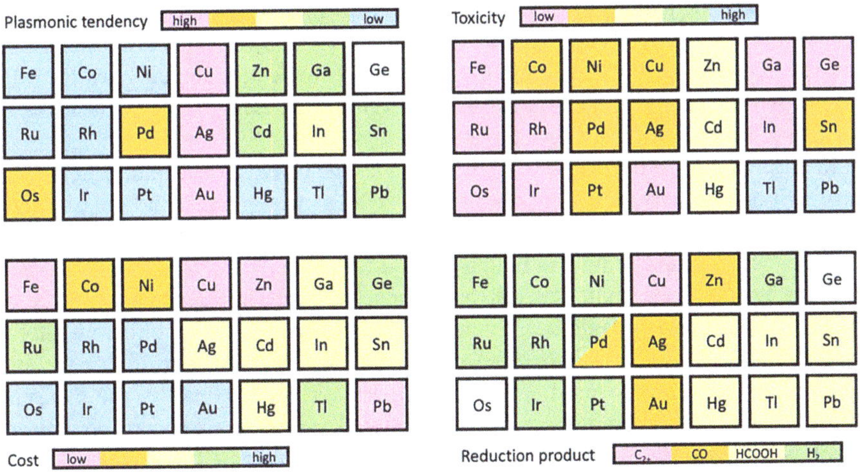

FIGURE 6.2 Consideration of material selection with four different concepts including plasmonic tendency, toxicity, cost, and reduction product.

These materials with their intrinsic surface properties can catalyze HER or CO_2R, with the main products being H_2, HCOOH, CO, and C_{2+} chemicals. It is important to note that Cu is the only element capable of reducing CO_2 and conducting C–C coupling for higher value product. Instead, several options can be chosen to generate CO, including Au, Ag, Zn, and Pd. But Pd also presents comparable H_2 generation as CO production. Similarly, one can select from Cd, In, Sn, Hg, Tl, and Pb if COOH is the main desired product. Combination and innovation with different elements and variant geometries have been developed to generate other products than the intrinsic one. Precise surface control will be the key to tune the chemical reaction.

In addition to the consideration of catalytic property, toxicity and cost are usually the main concern when it comes to selection of materials. Take generation of HCOOH for example, Tl and Pb would be the least favorable choice due to the high toxicity, while In and Sn would be the better candidates. As for CO production, Ag would be the best option. Furthermore, theories have predicted that d-band metals can be optically excited above interband with specific hot carrier distribution that depends on incident photon energy. Regarding plasmonic photocatalysis, higher plasmonic tendency is beneficial to harvest the sunlight for hot carrier generation. Cu, Ag, and Au outperform the rest and are suitable to perform CO_2R. Light management strategies have been applied to enable increased transparency of metal catalysts and enhanced absorption in semiconductors to generate more carriers for catalytic reaction. Usually, optical simulations are conducted to demonstrate the concept and optimize the geometry parameters.

Despite much promise, the main limitations of plasmonic photocatalysis are the insufficient light-matter interaction and difficulties associated with charge transfer, bringing about less than <0.5% incident photon-to-electron conversion efficiency (IPCE), leading to even lower energy efficiency after considering faradaic efficiency (FE, electron to chemical conversion efficiency). Besides, the capture of plasmonic hot holes presents much greater challenges than hot electrons, due to the fact of shorter mean-free paths for hot holes (5–10 nm, 1–2 eV below the Femi level) than hot electrons (20 nm, 1–2 eV above the Femi level) [2, 3]. Fundamental study and innovative design strategy to tackle these technical challenges and enhance the efficiency by order of magnitude needs to be developed to enable the practical use of these materials for solar-to-fuels applications.

Metasurface as an intriguing platform provides the power to manipulate light with desired properties, including but not limited to intensities, phases, and polarization. Coupling between different optical modes offers an additional degree of freedom to tune the electromagnetic field in both near- and far-fields. Hot carrier generation and dynamic in plasmonic materials can be influenced by the surrounding light-matter interaction with proper designs. The understanding of the coupling effect on photochemistry can also be applied to other more established PEC systems for enhancing product selectivity and overall efficiency. Besides energy application, the metasurface platform can be extended further for possible chiral sensing and separation, which is beneficial for the chemistry/biology society.

The simplest model of coupling arises from describing a single emitter in a cavity with the quantum picture as Jaynes-Cummings Hamiltonian: [4]

$$H_{JC}\Psi = \begin{bmatrix} \hbar\omega_{cav} & \hbar g \\ \hbar g & \hbar\omega_{em} \end{bmatrix} \begin{bmatrix} \Psi_{cav} \\ \Psi_{em} \end{bmatrix} = E \begin{bmatrix} \Psi_{cav} \\ \Psi_{em} \end{bmatrix}$$

$$\omega_{\pm} = \frac{\omega_{em} + \omega_{cav}}{2} \pm \frac{1}{2}\sqrt{(\omega_{cav} - \omega_{em})^2 + 4g^2} = \frac{\omega_{em} + \omega_{cav}}{2} \pm \Omega_R$$

where ω_{cav} indicates the cavity resonance frequency, ω_{em} indicates the emitter frequency, g is the coupling strength, and Ω_R defines the Rabi frequency. The maximum coupling strength is giving by $g = \frac{\mu_{em}}{\hbar}\sqrt{\frac{\hbar\omega_{cav}}{2V}}$, where V is the mode volume and μ_{em} is the dipole moment of emitter. Further account for damping term the normal-mode frequencies become:

$$\omega_{\pm} = \frac{\omega_{em} + \omega_{cav}}{2} - \frac{1}{4}i(\gamma_{em} + \gamma_{cav}) \pm \Omega_R$$

$$\Omega_R = \frac{1}{2}\sqrt{(\omega_{cav} - \omega_{em})^2 + g^2 - \frac{1}{4}(\gamma_{cav} - \gamma_{em})^2}$$

where γ_{cav} and γ_{em} indicate cavity and emitter decay rate accordingly. If considering n emitters, the Hamiltonian can be solved with eigenvalues split at resonance of $2\sqrt{n}\Omega_R$ [5]. The indistinguishable emitters would form a collective state. Therefore, the total Rabi splitting scales with square root of n.

There are two regimes of light-matter coupling defining with the coupling strength. For $g < \frac{1}{4}(\gamma_{cav} + \gamma_{em})$, the weak coupling would not present a significant resonance shift but show on-resonance enhancement and off-resonance suppression of the emission rate, also known as the Purcell effect. For $g > \frac{1}{4}(\gamma_{cav} + \gamma_{em})$, the strong coupling results in a significant shift of resonance between the lower branch and upper branch modes (Rabi splitting) and indicates formation of an exciton-photon hybrid state. The energy exchange between emitter and cavity photons is reversible. The requirements for reaching a strong coupling condition include strong field enhancement and overlap between field localizations. The high packing density of the plasmonic dipoles also improves the coupling strength. Several studies have successfully demonstrated strong-coupling-facilitated carrier transfer [6, 7]. The strong near-field enhancement can also influence the energy of the catalytic reaction, for example, strengthening CO binding to assist with C–C coupling for higher-value products like ethylene in CO_2R.

6.4 EXPERIMENTAL DEMONSTRATIONS

Photo-driven CO_2R with wired PV-EC configurations have shown great promise with high efficiency at bench scale operations. 19% solar-to-CO efficiency is realized by the combination of GaInP/GaInAs/Ge 3J photovoltaic, Ag gas diffusion electrode

(GDE) as cathode, and Ni foam as anode [8]. An overall efficiency of 5.6% to higher value products, including ethylene, ethanol and propanol, was achieved by integration of Cu/Ag catalysts with series-connected Si cells and power matching electronics [9]. The other approach utilizing Ni/InGaP/GaAs photoanode with a Pd/C/Ni cathode demonstrates a 10% solar-to-formate efficiency [10]. However, these prototypes are limited to dark electrolysis for cathodic CO_2R.

State-of-the-art design of integrated PEC for front-illuminated photocathodic CO_2R are listed and compared in Figure 6.3. To achieve a higher photon to product efficiency, consideration for shadow loss, photocurrent, and catalytic area need to be included. The properties of different categories among the 6 designed geometries are ranking from 1 to 6, with a lower value indicating better performance.

Nanostructured plasmonic metal as indicated in the first column of Figure 6.3 presents high catalytic surface area, which by itself functions as an absorber [11]. Therefore, no shadow loss need be a concern. However, the short carrier lifetime and low charge utilization result in the lowest photocurrent compared to other semiconductor-based light absorbers. The overall efficiency is then still limited to a non-practical level. Instead, an efficiency of 5% for CO_2 to CO conversion was achieved by use of prism-shaped Ag catalysts integrated on triple-junction III-V cells [12]. The concept is illustrated in the last column of Figure 6.3 where relatively high catalytic surface area and medium shadow loss result in comparable photocurrent and therefore the highest photon to product efficiency. Another approach includes dielectric nanoconical antenna arrays allowing incident light to couple into waveguide modes that enable light absorption in the photocathode below [13]. Low shadow loss and high photocurrent with relativity can be realized. But lower conversion is expected due to the slightly less catalytic surface. Both the concept of prism-shaped catalysts and a nanoconical dielectric allow opaque catalyst materials to be effectively transparent while still maintaining reasonable catalytic activity.

Geometry						
Shadow loss (1: lowest 6: highest)	1	2	6	5	3	4
Photocurrent (1: highest 6: lowest)	6	1	5	4	2	3
Catalytic area (1: highest 6: lowest)	1	6	4	5	3	2
Photon to Product Efficiency (1: highest 6: lowest)	6	5	4	3	2	1

FIGURE 6.3 Comparison of different designs of front-illuminated photocathode.

Previous demonstrations involved nanowire or microwire of semiconductors and nanoparticulated catalysts in the PEC system, as shown in the second column of Figure 6.3 [14]. It presents minimized shadow loss and the highest photocurrent, due to optimized absorption in the light absorber. Even though this is a success for HER since only a limited number of catalysts are required, the approach doesn't work well for CO_2R, due to the commonly lower catalytic activity for CO_2R where higher reaction surface area would be necessary. Some developments apply patches of metal catalysts to prevent complete blocking of light as exhibited in the third column of Figure 6.3 [15]. The semiconductor can extrude from the surface to enhance absorption, which is presented in the fourth column of Figure 6.3 [16]. However, the dramatic shadow loss still dominates and limits the photocurrent. The system is expected to show medium photon to product efficiency as compared to others.

A summary of recent progress with plasmonic photocatalysis is discussed. Products including CH_4, C_2H_6, CH_3OH, and HCHO can be generated when a UV light source is applied with higher energy to excite a d-band electron in plasmonic Au at 75°C [17]. Thermal catalytic reaction operated at elevated temperature can also be facilitated with light [18]. For example, Sabatier reaction which transforms CO_2 into CH_4 with addition of H_2 was proved to be enhanced with Au and Ru nanoparticles on Siralox® substrate (Au–Ru–S) where Au NP acts as a light absorber and Ru NP as the catalytic center [19]. The 'antenna-reactor' type photocatalyst with the Al@Cu_2O core–shell structure also showed great promise for reverse water-gas shift reaction, where CO is produced instead of CH_4 [20]. The photothermal effect and hot-carrier driven mechanistic pathways would need to be distinguished with careful control experiments.

To date, there are many plasmonic systems capable of harvesting hot electrons using plasmonic metal nanoparticles coupled to an n-type semiconductor for plasmon-driven photocatalysis, including photocatalytic hydrogen production, water oxidation, overall water splitting, and small molecules activation. However, only few studies focused on the transferring of plasmonic hot holes, which involves the capturing and transferring of plasmonic hot holes from the metal nanoparticles using wide bandgap p-type semiconductors. For photo-induced CO_2R to chemicals with electrical bias, it has been shown that plasmon-driven selectivity can be achieved with plasmonic Ag thin film [21].

To more efficiently separate the generated hot carriers, most of the studies have focused on the Au/TiO_2 involving the capture of hot electrons, but rarely focused on the capture of plasmonic hot holes by p-type semiconductors. Lately, the collection of hot-holes from Au NPs utilizing p-type gallium nitride is demonstrated and able to show photoelectrochemical CO_2R with such a plasmonic Au/p-GaN photocathode system [22]. A sufficient Schottky barrier over 1.0 eV was confirmed for facilitating the separation of hot carriers at the Au/p-GaN interface. Also, enhanced selectivity for photoelectrochemical CO_2 reduction to CO and COOH over hydrogen evolution using a plasmonic Cu/p-NiO photocathode is observed [23]. Nevertheless, these photocathode devices were realized in aqueous electrolytes under the applied potential (~ −1.0 V vs. RHE), which provides the additional driven force to quickly migrate the charge carriers once separated. More recently, a report has shown the employment of plasmonic devices to capture hot holes for achieving zero-biased photocatalytic CO_2

reduction at room temperature and visible illumination, especially using H_2O molecules as electron donors and without any sacrificial reagents [24].

Regarding enhancing light-matter interaction, there are simulation efforts that reach strong coupling conditions between SPR and dielectric resonance. Experimental results present coupling between Fabry Pérot nanocavity mode and the localized SPR with Au NPs/TiO_2/Au film structure [25]. This is also the first demonstration of photocatalytic performance enhancement that directly correlates with a strong coupling condition. With the presence of a sacrificial electron donor, the IQE can reach 3% at visible wavelength [26].

6.5 CONCLUSION AND OUTLOOK

We believe, with proper design based on optical simulation and coupling with stronger dielectric resonance, a high efficiency system can be realized without a sacrificial reagent. The hybrid system also reveals the additional possibility to study more challenging and complicated reduction reactions under a strong coupling regime. Also, the innovative concept can be extended to other chemical reactions. However, showing potential with a laboratorial scale demonstration is never enough. Further efforts and developments by the research community would be required to realize a practical application of photo-driven CO_2 reduction. Equilibrium between efficiency, scalability, and stability need to be considered to bring the solar fuels system to market [27].

REFERENCES

[1] Cheng W-H, Richter MH, May MM, et al. Monolithic photoelectrochemical device for direct water splitting with 19% efficiency. *ACS Energy Lett* 2018; 3: 1795–1800.

[2] Bernardi M, Mustafa J, Neaton JB, et al. Theory and computation of hot carriers generated by surface plasmon polaritons in noble metals. *Nat Commun* 2015; 6: 1–9.

[3] Tagliabue G, DuChene JS, Abdellah M, et al. Ultrafast hot-hole injection modifies hot-electron dynamics in Au/p-GaN heterostructures. *Nat Mater* 2020; 19: 1312–1318.

[4] Pelton M, Storm SD, Leng H. Strong coupling of emitters to single plasmonic nanoparticles: exciton-induced transparency and Rabi splitting. *Nanoscale* 2019; 11: 14540–14552.

[5] Vasa P. Exciton-surface plasmon polariton interactions. *Adv Phys X* 2020; 5: 1749884.

[6] Hertzog M, Wang M, Mony J, et al. Strong light-matter interactions: a new direction within chemistry. *Chem Soc Rev* 2019; 48: 937–961.

[7] Shan H, Yu Y, Wang X, et al. Direct observation of ultrafast plasmonic hot electron transfer in the strong coupling regime. *Light Sci Appl* 2019; 8: 1–9.

[8] Cheng W-H, Richter MH, Sullivan I, et al. CO_2 reduction to CO with 19% efficiency in a solar-driven gas diffusion electrode flow cell under outdoor solar illumination. *ACS Energy Lett* 2020; 29: 470–476.

[9] Gurudayal, Bullock J, Srankó DF, et al. Efficient solar-driven electrochemical CO_2 reduction to hydrocarbons and oxygenates. *Energy Environ Sci* 2017; 10: 2222–2230.

[10] Zhou X, Liu R, Sun K, et al. Solar-driven reduction of 1 atm of CO_2 to formate at 10% energy-conversion efficiency by use of a TiO_2-protected III-V tandem photoanode in conjunction with a bipolar membrane and a Pd/C cathode. *ACS Energy Lett* 2016; 1: 764–770.

[11] Welch AJ, Duchene JS, Tagliabue G, et al. Nanoporous gold as a highly selective and active carbon dioxide reduction catalyst. *ACS Appl Energy Mater* 2019; 2: 164–170.

[12] Cheng W-H, Richter MH, Müller R, et al. Integrated solar-driven device with a front surface semitransparent catalysts for unassisted CO_2 reduction. *Adv Energy Mater* 2022; 12: 2201062.

[13] Yalamanchili S, Verlage E, Cheng W-H, et al. High broadband light transmission for solar fuels production using dielectric optical waveguides in TiO_2 nanocone arrays. *Nano Lett* 2020; 20: 502–508.

[14] Kempler PA, Richter MH, Cheng W-H, et al. Si microwire-array photocathodes decorated with Cu allow CO_2 reduction with minimal parasitic absorption of sunlight. *ACS Energy Lett* 2020; 5: 2528–2534.

[15] Tae Song J, Ryoo H, Cho M, et al. Nanoporous Au thin films on Si photoelectrodes for selective and efficient photoelectrochemical CO_2 reduction. *Adv Energy Mater* 2017; 7: 1601103.

[16] Narasimhan VK, Hymel TM, Lai RA, et al. Hybrid metal-semiconductor nanostructure for ultrahigh optical absorption and low electrical resistance at optoelectronic interfaces. *ACS Nano* 2015; 9: 10590–10597.

[17] Hou W, Hung WH, Pavaskar P, et al. Photocatalytic conversion of CO_2 to hydrocarbon fuels via plasmon-enhanced absorption and metallic interband transitions. *ACS Catal* 2011; 1: 929–936.

[18] Zhang X, Li X, Zhang D, et al. Product selectivity in plasmonic photocatalysis for carbon dioxide hydrogenation. *Nat Commun* 2017; 8: 1–9.

[19] Mateo D, De Masi D, Albero J, et al. Synergism of Au and Ru nanoparticles in low-temperature photoassisted CO_2 methanation. *Chem – A Eur J* 2018; 24: 18436–18443.

[20] Robatjazi H, Zhao H, Swearer DF, et al. Plasmon-induced selective carbon dioxide conversion on earth-abundant aluminum-cuprous oxide antenna-reactor nanoparticles. *Nat Commun* 2017; 8: 1–10.

[21] Creel EB, Corson ER, Eichhorn J, et al. Directing selectivity of electrochemical carbon dioxide reduction using plasmonics. *ACS Energy Lett* 2019; 4: 1098–1105.

[22] DuChene J S, Tagliabue G, et al. Hot hole collection and photoelectrochemical CO_2 reduction with plasmonic Au/p-GaN photocathodes. *Nano Lett* 2018; 18: 2545–2550.

[23] Duchene JS, Tagliabue G, Welch AJ, et al. Optical excitation of a nanoparticle Cu/p-NiO photocathode improves reaction selectivity for CO_2 reduction in aqueous electrolytes. *Nano Lett* 2020; 20: 2348–2358.

[24] Li R, Cheng W-H, Richter MH, et al. Unassisted highly selective gas-phase CO_2 reduction with a plasmonic Au/p-GaN photocatalyst using H_2O as an electron donor. *ACS Energy Lett* 2021; 6: 1849–1856.

[25] Shi X, Ueno K, Oshikiri T, et al. Enhanced water splitting under modal strong coupling conditions. *Nat Nanotechnol* 2018; 13: 953–958.

[26] Cao Y, Oshikiri T, Shi X, et al. Efficient hot-electron transfer under modal strong coupling conditions with sacrificial electron donors. *ChemNanoMat* 2019; 5: 1008–1014.

[27] Segev G, Kibsgaard J, Hahn C, et al. The 2022 solar fuels roadmap. *J Phys D Appl Phys* 2022; 55: 323003.

7 Water Chestnut Shell-Derived Carbons as Electrodes for Energy Applications

Han-Yi Chen
National Tsing Hua University, Hsinchu City, Taiwan

7.1 INTRODUCTION

Over the past few years, there has been a growing interest in the utilization of biomass-derived materials for various applications in energy storage and power generation [1]. Among these sustainable resources, water chestnut shells (also called water caltrop shells or *Trapa natans* husks) are favorable precursors for the synthesis of carbon-based materials with unique properties [2]. Water chestnuts, a widely cultivated fruit crop in Asia, yield more than 3400 tons of husk waste annually in Taiwan alone [3]. Unfortunately, the conventional approach of burning water chestnut husk waste performed by local farmers contributes significantly to air pollution. Consequently, repurposing this waste by converting it into cost-effective carbon electrode materials has the potential to establish a sustainable circular economy [4]. This chapter explores the versatile applications of carbon derived from these abundant agricultural byproducts, focusing on their pivotal role as low-cost and sustainable electrode materials [1, 3, 5–8].

The following chapters discuss the synthesis methods and characterization of different water chestnut shell-derived carbons. These carbon materials exhibit remarkable potential for use in high-performance supercapacitors [1, 3], sodium-ion batteries [6], and potassium-ion battery electrodes [5], thus offering a sustainable alternative to conventional energy storage systems. Beyond energy storage applications, this chapter also explores the use of water chestnut shell-derived nanoporous carbons in sustainable high-power microbial fuel cells [7] and plant microbial fuel cells.[8] These materials present exciting possibilities for environmentally friendly power generation.

This chapter provides an overview of the diverse applications of carbon materials derived from water chestnut shells, highlighting their role as versatile and sustainable electrode materials in a range of energy storage and power generation technologies. Through the exploration of these innovative materials, we can contribute to the development of cleaner and more efficient energy solutions, while simultaneously addressing environmental challenges.

DOI: 10.1201/9781003463009-7

7.2 PREPARATION METHODS AND CHARACTERIZATION OF WATER CHESTNUT HUSKS-DERIVED CARBONS

This section provides a comprehensive overview of the various synthetic methods employed to produce carbon materials from water chestnut shells. Typically, biomass-derived carbon is prepared via carbonization and activation. The following sections detail the procedures for the synthesis of biomass-derived carbon.

7.2.1 Preparation of Biomass-Derived Carbons

7.2.1.1 Carbonization

Generally, precursor materials are heated in an inert gas environment for carbonization. Within the temperature range of 200–500 °C, initial thermal decomposition occurs within the precursor material, with organic compounds other than carbon being removed through volatilization and tar formation. Carbon-containing bonds undergo condensation. When the temperature exceeds 500 °C, carbonization reactions begin, and the initial char formed from thermal decomposition has an sp^3 structure. The condensation reactions continue to form graphite-like microcrystals with sp^2 structures, resulting in the formation of irregular porous carbon structures. As the carbonization temperature increases, the degree of graphitization improves, making the graphitized structure more complete; and the microcrystal arrangement becomes more orderly, resulting in better conductivity. Acid washing is used to remove byproducts, such as tar, produced during carbonization to prevent pore blockage, which reduces the surface area and porosity [9–11].

7.2.1.2 Activation

Activation can be carried out in two ways, physical or chemical.

Physical Activation: During carbonization of materials such as wood or coconut shells, internal carbon dioxide and water vapor serve as activators. This occurs at temperatures ranging from 800 °C to 1000 °C, resulting in the generation of pores within the material. The reaction equations are as follows: [12, 13]

$$\text{Reaction 1} : C + 2CO_2 \rightarrow 2CO, \Delta H = 159 \text{ kJ mol}^{-1} \tag{7.1}$$

$$\text{Reaction 2} : C + H_2O \rightarrow CO + H_2, \Delta H = 117 \text{ kJ mol}^{-1} \tag{7.2}$$

Chemical Activation: Chemical agents, typically those with dehydrating properties, are used as activators. Common activators include $ZnCl_2$ [14], NaOH [15], H_3PO_4 [16], and KOH [17]. Chemical activation offers two advantages over physical activation. It can be performed at lower temperatures, resulting in higher carbon yields. For instance, when zinc chloride is used as an activator, it degrades cellulose and promotes the dehydration and carbonization of materials at high temperatures, creating a porous structure. Additionally, thermal decomposition can inhibit tar formation and reduce the generation of volatile substances, thereby increasing the carbon yield. Recently, iron chloride has been recognized as an effective activator. It not only induces material dehydration during high-temperature carbonization but also facilitates material graphitization. During heating, Fe_3O_4 or Fe_2O_3 is generated, promoting pore formation [18, 19].

7.2.2 Preparation of Water Chestnut Shell-Derived Carbons

Several methods have been used to synthesize water chestnut shell-derived carbon [5, 6]. As the first step, the water chestnut shell is generally washed with water, ethanol, or HCl, then smashed into powder or cut into pieces. Then the powder is carbonized at high temperatures (200–1500 °C) under N_2 or Ar flow. After carbonization, the powder is cleaned with 1 M HCl and deionized water to eliminate any impurities. The procedure for the synthesis of water chestnut shell-derived carbon is shown in Figure 7.1 [5].

Various activation agents have been used to synthesize water chestnut shell-derived porous carbons, including ZnO [3], KOH [3], and $ZnCl_2$ [6]. Lin and Hsu's group mixed a carbon precursor (water caltrop shell (WCS) biochar) with ZnO nanoparticles and KOH using a blender, then enclosed it in a stainless-steel container and subjected it to heating through a top-lit-updraft technique using an appropriate carbonization furnace at 900 °C for 2 hours. The sample was then washed with HCl and water to remove the residual activation agents [3]. The formation mechanism and carbonization furnace are shown in Figures 7.2(a, b). The scanning electron microscopy (SEM) images (Figures 7.2(c, d)) and N_2 adsorption–desorption isotherms (Figure 7.2(e), curve I) show that the WCS biochar has many macropores with a specific surface area of 230 $m^2\ g^{-1}$. After activation, the surface area of the resulting WCS porous carbon is 1537 $m^2\ g^{-1}$ when the WCS biochar/ZnO/KOH weight ratio is 1.0:2.0:0.80 (Figure 7.2(e), curve II). The SEM (Figure 7.2 (f)) and resolution transmission electron microscopy (TEM) images (Figures 7.2(g, h)) show that the WCS porous carbon clearly exhibits a porous texture with mesopores and micropores. Chen and Liu's group prepared the *T. natans* (TNH) husk-derived nanoporous carbons by soaking the carbon precursor in a saturated $ZnCl_2$ solution for 24 hours at a

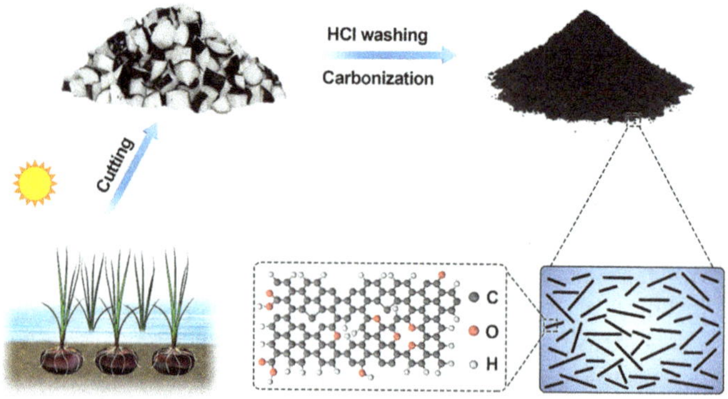

FIGURE 7.1 Synthesis of water chestnut-derived carbon by carbonization [5]. **(Reprinted with permission, copyright 2020, American Chemical Society.)**

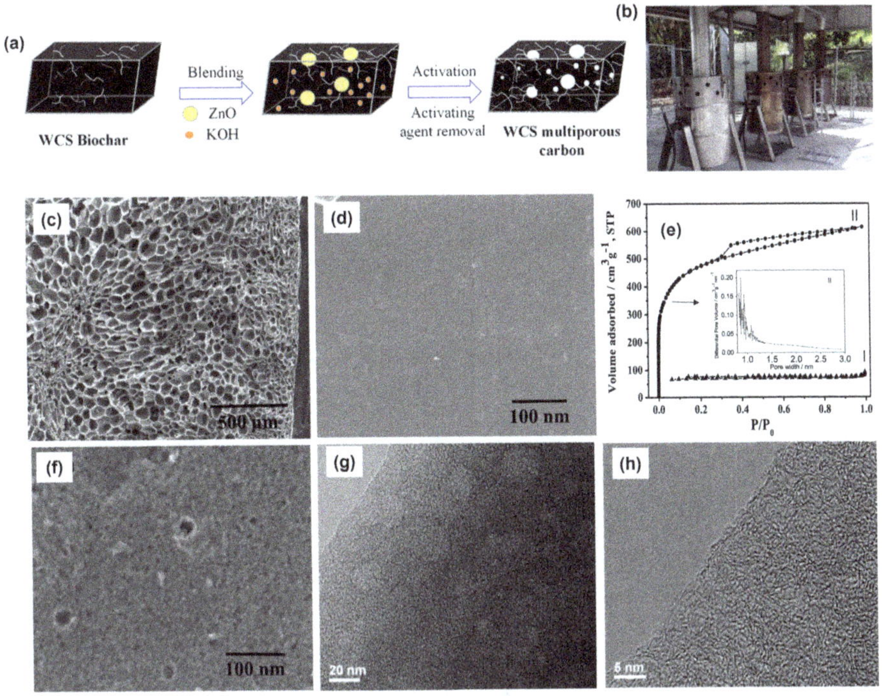

FIGURE 7.2 (a) Diagram depicting how mesopores and micropores are formed in water caltrop shell (WCS) biochar through the utilization of ZnO nanoparticles and KOH as activating agents; (b) carbonization furnace systems in the Guantian District, Tainan, Taiwan; (c, d) scanning electron microscopy images of the WCS biochar; (e) N_2 sorption behavior comparison between the WCS biochar (I) and the WCS multiporous carbon (II), with an accompanying pore size distribution graph provided in the inset; (f) scanning electron microscopy images; and (g, h) high-resolution transmission electron microscopy images of the WCS multiporous carbon [3].

weight ratio of 1:1, then heated in an N_2 atmosphere between 600 °C and 1000 °C for 2 hours [7]. The carbon powder obtained was cleaned with HCl and water. The obtained TNHs also delivered a high surface area of approximately 1500 $m^2\ g^{-1}$.

Nitrogen- and sulfur-doped water caltrop shell-derived porous carbons have proven to be promising electrode materials for CO_2 capture and supercapacitor applications. Hu and Xu's group synthesized N-doped water caltrop shell-derived porous carbon via a three-step procedure [20]. First, the carbon precursor was carbonized at 500 °C for 1 hour under N_2 atmosphere; then the powder was mixed with melamine with a mass ratio of 1:1 at 500 °C for 2 hours under N_2 atmosphere. The N-enriched powder then underwent KOH activation at temperatures of 500–650 °C. After washing and drying, the obtained carbon exhibited a high surface area of up to 2384 $m^2\ g^{-1}$ and N content of 8.48 wt %. Hu and Wang's group also used the same method to

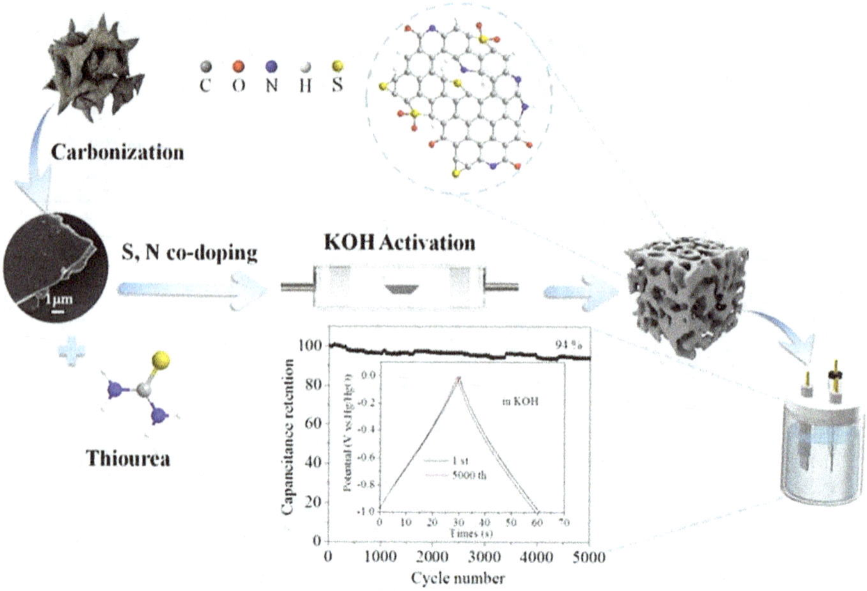

FIGURE 7.3 Diagram illustrating the production process of porous carbons doped with nitrogen and sulfur using the KOH activation method, where water chestnut shells serve as the carbon source and thiourea as the nitrogen and sulfur supplier [21].

prepare N/S co-doped water caltrop shell-derived porous carbons using thiourea as both the N and S supply reagent, as shown in Figure 7.3 [21]. The XPS results confirmed that the water caltrop shell-derived porous carbons were successfully doped with N (>3.55 wt %) and S (>0.49 wt %) atoms. The N- and S-containing functional groups on the carbon surface varied its electron density, which enhanced its electronic conductivity, wettability, and activity [7, 22].

7.3 ENERGY STORAGE APPLICATIONS: SUPERCAPACITORS AND BATTERIES

7.3.1 Supercapacitors

Supercapacitors, also known as electrochemical capacitors, have been widely utilized in energy storage applications owing to their high-power density, fast charge/discharge rate, and long cycle life [23, 24]. They can be used for portable electronics, transportation, memory backup systems, and uninterruptible power sources [23, 24]. In recent decades, porous carbons have garnered significant attention because of their exceptional properties, which make them highly promising for supercapacitor applications. Sustainable energy storage technology is achievable by producing environmentally friendly and affordable porous carbon doped with heteroatoms sourced from renewable biomass materials. Hu and Wang's group successfully synthesized N/S-doped water chestnut shell-derived porous carbon using the KOH activation method with thiourea as the nitrogen/sulfur source [21]. The resulting carbons exhibit high porosity (>1350 $m^2 g^{-1}$) with nitrogen- and sulfur-containing functional groups. These porous carbons were used as supercapacitor electrode materials and exhibited excellent electrochemical performance; the electrode achieved a high specific capacitance of 318 F g^{-1} at a current density of 0.5 A g^{-1}, along with outstanding cycling life, retaining 94% of its capacity after 5000 cycles.

Lin and Hsu's group introduced an economical, sustainable, and environmentally friendly approach for synthesizing water caltrop shell-derived porous carbons using a top-lit-updraft method (see Figure 7.2) with ZnO nanoparticles and KOH as activation agents, resulting in microporous carbon with a high surface area of up to 1537 $m^2 g^{-1}$ [3]. These porous carbon electrodes also demonstrated good supercapacitor performance, with a high specific capacitance of 128 F g^{-1} at 5 mV s^{-1} (Figures 7.4(a, b)). They exhibited good retention rates even at a high scan rate of 500 mV s^{-1} (>60% compared to the capacitance at 5 mV s^{-1}) and low ohmic resistance when tested in a 1.0 M $LiClO_4$/PC electrolyte, resulting in excellent cycling retention (>99%) after 10,000 cycles at a scan rate of 100 mV s^{-1} (Figure 7.4(c)). The galvanostatic charge/discharge test exhibited a high-power density of 6564 W kg^{-1} at 4.35 W h kg^{-1} energy density (Figures 7.4(d, e)). Furthermore, a 1500 F-class pouch cell-type supercapacitor with water caltrop shell-derived porous carbon electrodes was assembled to evaluate their electrochemical performance in practical applications (Figure 7.4(f)).

7.3.2 Sodium-Ion Batteries

Lithium-ion batteries (LIBs) are one of the most extensively used technologies in the energy storage field, especially for portable and electrical devices, because of their high energy density [25]. However, the scarcity of lithium in the Earth's crust increases the cost of lithium-ion batteries. Currently, sodium-ion batteries are potentially better alternative candidates for energy storage because of their low cost and the abundance of raw materials [26, 27]. However, the higher standard reduction potential of Na^+/Na than that of Li^+/Li and the heavier atomic mass and larger ionic radius of Na^+ result in lower energy density in sodium-ion batteries than in lithium-ion batteries. Despite these disadvantages, sodium-ion batteries still play a significant role in large-scale energy storage applications, owing to their cost advantages over lithium-ion batteries.

Carbon is a promising anode material for sodium-ion batteries because it is cheap, easily obtainable, and environmentally friendly [28]. Graphite is the most commonly used anode material for commercial lithium-ion batteries, but it is difficult to intercalate sodium ions reversibly into graphite with commonly used electrolytes because of unfavorable thermodynamic processes. Recently, hard carbon has been proposed as the most promising anode material for sodium-ion batteries; however, its high cost and low initial coulombic efficiency restrict its use in large-scale applications.

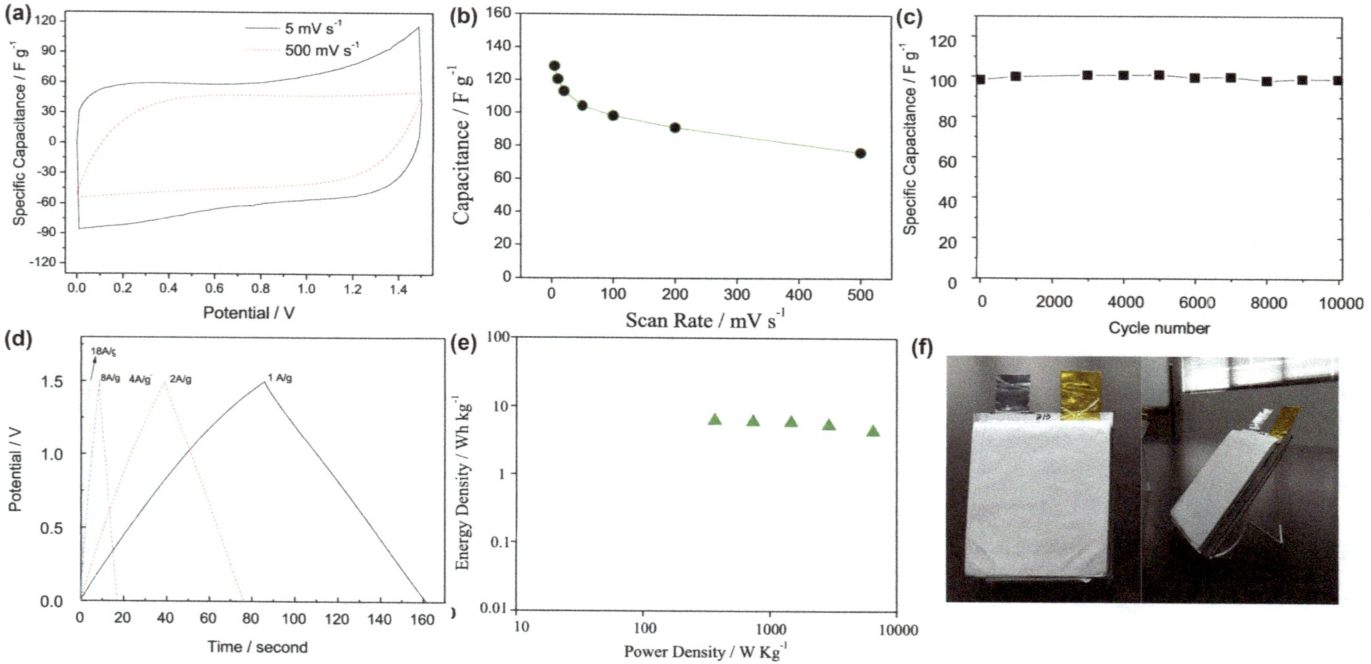

FIGURE 7.4 (a) Cyclic voltammogram curves at scan rates of 5 and 500 mV s^{-1}; (b) specific capacitance at different scan rates; and (c) cycling performance at a scan rate of 100 mV s^{-1} for the water caltrop shell-derived porous carbon electrodes; (d) galvanostatic charge-discharge curves under different current densities; (e) Ragone plot for the water caltrop shell-derived porous carbon electrodes in a 1.0 M $LiClO_4$/PC electrolyte; (f) 1500 F class pouch cell-typed water caltrop shell-derived porous carbon supercapacitor. (Length/width/height = 120 mm/100 mm/10 mm) [3].

(Reprinted with permission, copyright 2020, American Chemical Society.)

Fan and Shi introduced a novel approach utilizing water caltrop shells to produce hard carbons through one-step carbonation followed by acid treatment [6]. The water caltrop shell-derived hard carbon, obtained by carbonizing at 1300 °C, exhibits remarkable sodium-ion battery performance, as shown in Figure 7.5 [6]. At a current density of 0.1 C, the electrode demonstrates a reversible specific capacity of 394.0 mAh g^{-1}, with a plateau capacity as high as 285.2 mAh g^{-1} and an impressive initial coulombic efficiency of up to 84%. Even at a higher current density of 0.4 C, the reversible specific capacity remains substantial at 313.9 mAh g^{-1}, with only a minimal ~8% degradation observed after 200 cycles. This paper presents a promising biowaste-derived anode for high-energy sodium-ion batteries.

7.3.3 Potassium-Ion Batteries

Potassium-ion batteries are also considered a potential alternative to lithium-ion batteries because of their affordability, abundance, non-toxicity, and low redox potential ($K/K^+ = -2.925$ V vs. standard hydrogen electrode) [29]. To date, various efforts have been dedicated to the development of new anode materials for potassium-ion batteries. However, the main obstacle faced by potassium-ion batteries is the larger ionic radius of potassium ions (1.38 Å) compared to those of the lithium ions (0.76 Å) or sodium ions (1.02 Å). This difference often limits the potassium ions to insert into anode materials, which leads to lower capacity, rate performance, and cycling stability under the charging/discharging processes [30]. High-surface-area disordered carbons exhibit stable chemical properties, large interlayer spacing, and high electronic conductivity, thus they have been recognized as suitable anode materials for potassium-ion batteries [31]. These properties facilitate the accommodation of more K^+ ions and mitigate volume changes. Heteroatom doping is a useful method for modifying the electronic structure and interlayer spacing of carbon materials [32]. Nevertheless, the design and manufacture of porous, heteroatom-doped, disordered carbon materials with high surface areas to enhance the potassium storage performance remains a significant challenge.

Zhou's group proposed a carbon anode derived from water chestnut shells, produced under moderate-temperature carbonization at 900 °C, for potassium-ion batteries [5]. This carbon anode exhibited a high initial reversible capacity of 253 mAh g^{-1} at 100 mA g^{-1}, with an excellent cycling performance of 220 mAh g^{-1} after 1000 cycles (Figures 7.6(a, b, e)). It also exhibited excellent rate capability of 135 mAh g^{-1} at 1000 mA g^{-1} (Figures 7.6(c, d)). When combined with a potassiated 3,4,9,10-pery lene-tetracarboxylic dianhydride cathode, the first capacity reached 124 mAh g^{-1} at 25 mA g^{-1}, with a capacity retention of 84.7% after 300 cycles at 50 mA g^{-1}. The enhanced capability of these materials to store potassium can be attributed to their disordered microstructure, which is characterized by a relatively high concentration of oxygen-related defects, a substantial specific surface area, and the presence of randomly-distributed short graphene nanosheets. Density functional theory calculations revealed that oxygen doping altered the charge density distribution of carbon effectively and promoted K-ion adsorption on water chestnut-derived carbon, thereby enhancing the capability of K-ion storage.

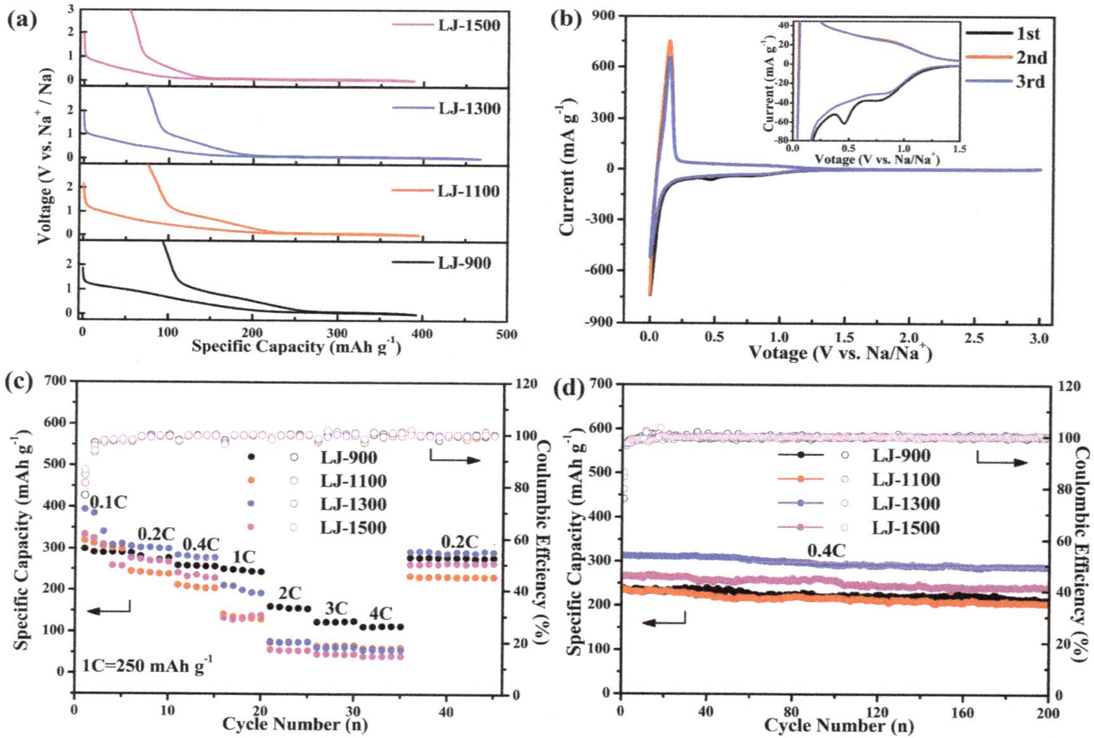

FIGURE 7.5 Sodium-ion half-cell test results of water caltrop shell-derived hard carbon electrodes (LJ-x, where x represents carbonization temperature): (a) galvanostatic 1st discharge/charge profiles; (b) CV curves of LJ-1300; (c) rate performance; (d) cyclic performance of at 0.4 C [6].

(Reprinted with permission, copyright 2019, Elsevier.)

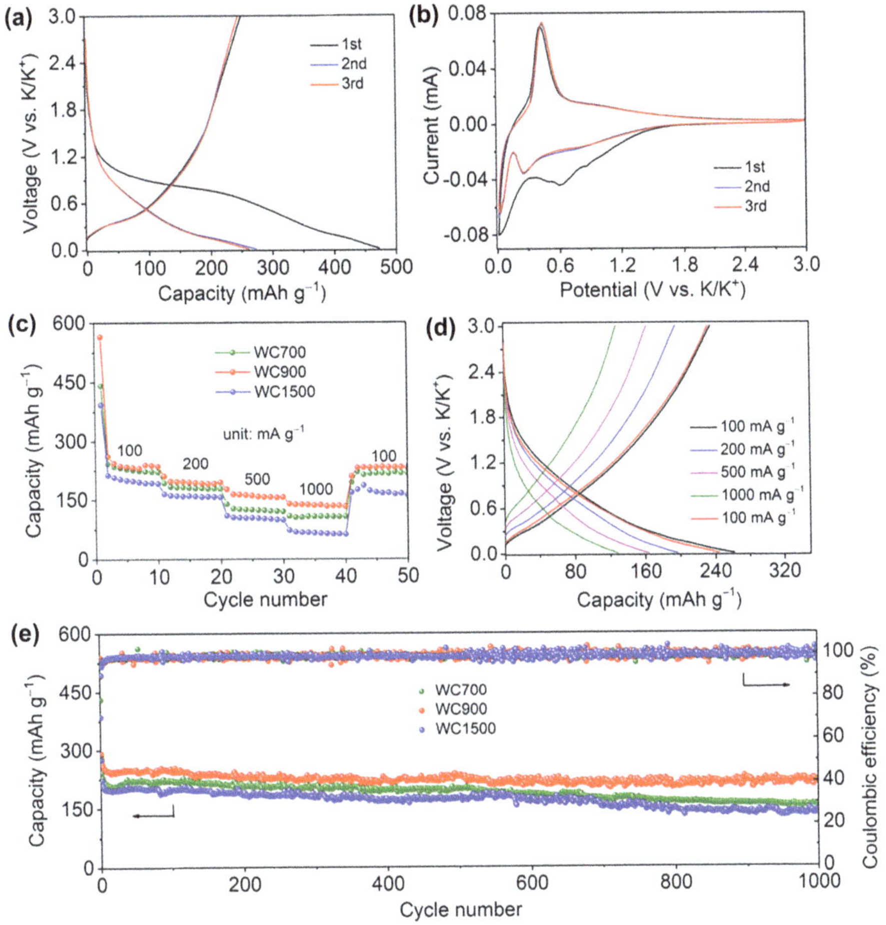

FIGURE 7.6 Battery performance of water chestnut shell-derived carbon electrodes carbonized at 700, 900, and 1500 °C (denoted as WC700, WC900, and WC1500): (a) charge–discharge curves at a current density of 100 mA g^{-1} and (b) cyclic voltammetry curves at a scan rate of 0.1 mV s^{-1} in the initial three cycles of the WC900 electrode; (c) rate capability of WC700, WC900, and WC1500; (d) charge–discharge curves of WC900 at different current densities; (e) long cycling test of WC700, WC900, and WC1500 at 100 mA g^{-1} [5].

7.4 APPLICATIONS IN POWER GENERATION: MICROBIAL AND PLANT MICROBIAL FUEL CELLS

7.4.1 Microbial Fuel Cells

Microbial Fuel Cells are a promising sustainable energy technology that converts chemical energy into electricity through microbial activity (Figure 7.7(a)). However, they face challenges related to high cost and low power output. Chen and Liu's group

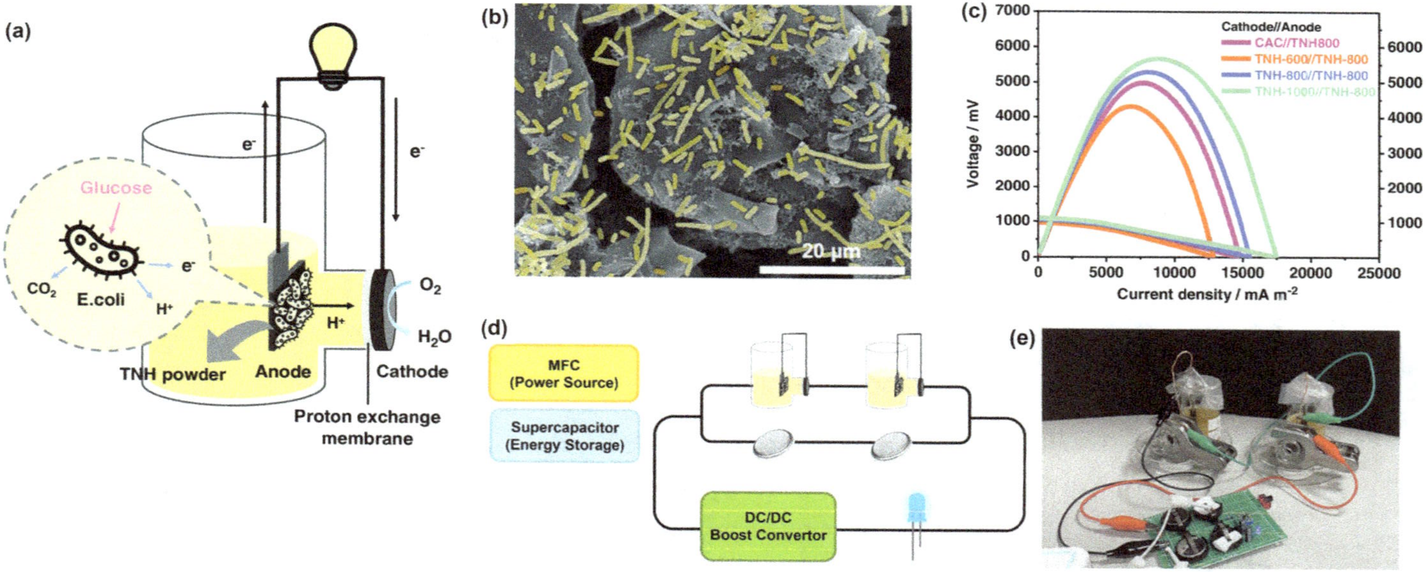

FIGURE 7.7 (a) Microbial fuel cell construction; (b) scanning electronic microscopy images of *E. coli* adhesion on *T. natans* husk (TNH)-derived nanoporous carbon electrode carbonized at 800 °C; (c) polarization and power density curves of different TNH-derived nanoporous carbon electrodes in the microbial fuel cells; (d) schematic; and (e) photograph of the microbial fuel cell-supercapacitor-LED system [7].

tried to overcome these issues by recycling agricultural waste, *T. natans* husks, to produce cost-effective nanoporous carbon. *T. natans* husk (TNH)-derived nanoporous carbons have been utilized as electrode materials in microbial fuel cells with *Escherichia coli* [7].

The TNH-derived nanoporous carbons were activated by $ZnCl_2$ (1:1) and carbonization at 600–1000 °C. After optimization, the material carbonized at 800 °C (surface area: 1324 m^2 g^{-1}) served as an excellent anode material with enhanced *E. coli* adhesion (Figure 7.7(b)) and conductivity; while the material carbonized at 1000 °C (surface area: 1375 m^2 g^{-1}) acted as a cathode material with improved oxygen reduction reaction activity. An impressive average power density of 5713 mW m^{-2} was achieved, surpassing commercial activated carbon (2998 mW m^{-2}) by 1.9 times (Figure 7.7(c)). TNH-derived nanoporous carbons exhibited superior bacterial adhesion and electrochemical activity owing to their favorable pore size distribution, suitable oxygen- and nitrogen-containing functional groups, high surface area, excellent biocompatibility, and conductivity.

Furthermore, this study explored the utilization of supercapacitors with TNH-derived nanoporous carbon electrodes to store the energy generated by the microbial fuel cells. The supercapacitors utilizing the carbon electrode carbonized at 600 °C (surface area: 1499 m^2 g^{-1}) demonstrated a remarkable specific capacitance of 84 F g^{-1} at a current density of 1 A g^{-1}, even after 1000 cycles in 1 M Na_2SO_4 in a two-electrode configuration. This configuration successfully lit up an LED for 15 min, demonstrating the practicality of a microbial fuel cell–supercapacitor system with TNH-derived nanoporous carbon electrodes (Figures 7.7(d, e)). This research underscores the economic value of using TNHs as raw materials for microbial fuel cells and supercapacitor electrodes, while simultaneously reducing CO_2 emissions associated with waste disposal. TNH-derived nanoporous carbon electrodes show significant promise as sustainable energy sources for advancing microbial fuel cells.

7.4.2 Plant Microbial Fuel Cells

Plant microbial fuel cells are a developing technology. They use organic substances released by plant roots to supply nutrients to microbes in the rhizosphere, thereby producing electricity, as shown in Figure 7.8(a) [8]. Plant microbial fuel cells have been integrated into various applications, such as soil remediation and greenhouse gas reduction [33]. They have also been used in constructed wetland plant microbial fuel cells, rooftop greenhouses, and as power sources for sensors and wireless communication devices [4]. Consequently, plant microbial fuel cells show significant promise for future advancement. However, their limited power density has hindered the advancement of plant microbial fuel cells.

Chen and Liu's group introduced TNH-derived carbon as a promising and sustainable electrode material for *Canna indica*-based plant microbial fuel cells [8]. The TNH-derived carbon used in this study was obtained from Guantian Black Gold, a Taiwanese company, using a top-lit-updraft approach with an appropriate carbonization furnace, as illustrated in Figure 7.2(b). Notably, this setup resulted in minimal carbon dioxide emissions, thereby mitigating environmental pollution. As husk waste has no intrinsic market value, the cost of TNH-derived carbon is roughly

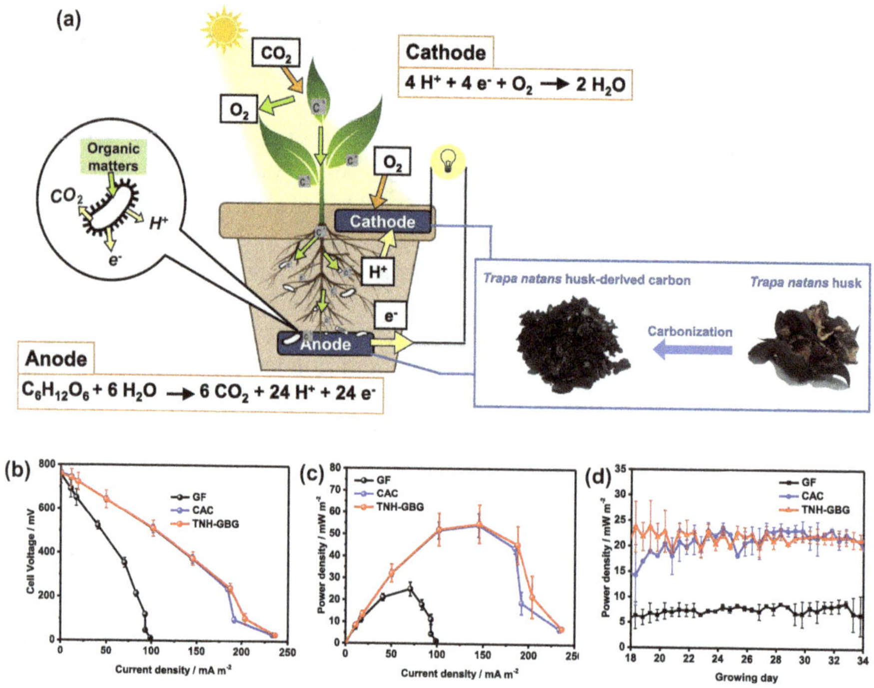

FIGURE 7.8 (a) Mechanisms of plant microbial fuel cell with the *T. natans* husk (TNH)-derived carbon electrode; (b) polarization curves and (c) power densities of plant microbial fuel cells with graphite felt, commercially activated carbon, or TNH-derived carbon electrodes obtained by load voltage measurements at day 40; (d) long-term power density measurements of plant microbial fuel cells connected to a 1000 Ω external load in series [8].

(Reprinted with permission, copyright 2022, Elsevier.)

one-fifth that of commercially available activated carbon. Consequently, repurposing this husk waste into affordable plant microbial fuel cell carbon electrode materials will contribute to establishing a sustainable circular economy. A maximum power density of 55 mW m^{-2} can be obtained from the plant microbial fuel cells with TNH-derived carbon-coated graphite felt electrodes via polarization curve measurements, which exhibited a significant improvement compared to that exhibited by the plant microbial fuel cells with pure graphite felt electrodes (22 mW m^{-2}), as shown in Figures 7.8(b, c). The increased power density of the TNH-derived carbon-coated graphite felt electrodes can be attributed to their substantial surface area and rich oxygen-containing functional groups, which enhance hydrophilicity and likely promote microbial attachment, thus reducing activation polarization. Notably, TNH-derived carbon-based plant microbial fuel cells exhibited a performance like that of commercially activated carbon-coated graphite felt electrodes, but at only one-fifth of the cost (Figure 7.8(d)). This study demonstrates the potential of TNH-derived carbon as a cost-effective electrode material for sustainable power generation applications.

7.5 SUMMARY

This chapter explores the diverse applications of carbon materials derived from water chestnut shells, focusing on their pivotal role as low-cost and sustainable electrode materials. These materials have demonstrated significant potential in various fields.

Energy Storage Applications: Water chestnut shell-derived carbons have remarkable potential for use in high-performance supercapacitors, sodium-ion batteries, and potassium-ion batteries. They offer high capacity and excellent cycling stability.

Power Generation Applications: Water chestnut shell-derived nanoporous carbon has been integrated into fuel cells and plant fuel cells. The cells exhibit excellent power generation capabilities and bacterial adhesion, thus providing sustainable energy sources with potential applications in soil remediation, greenhouse gas reduction, and power supply for various devices.

Overall, this study highlighted the versatility and sustainability of carbon materials derived from water chestnut shells. Their application in energy storage and power generation technologies offers the potential for cleaner and more efficient energy solutions while addressing environmental challenges. These innovative materials can contribute to the development of sustainable and circular economies.

REFERENCES

[1] R. Chakraborty, Vilya K, M. Pradhan, A.K. Nayak, Recent advancement of biomass-derived porous carbon based materials for energy and environmental remediation applications, *Journal of Materials Chemistry A*, 10 (2022) 6965–7005.

[2] L. Rao, S. Liu, L. Wang, C. Ma, J. Wu, L. An, X. Hu, N-doped porous carbons from low-temperature and single-step sodium amide activation of carbonized water chestnut shell with excellent CO_2 capture performance, *Chemical Engineering Journal*, 359 (2019) 428–435.

[3] C.-H. Hsu, Z.-B. Pan, C.-R. Chen, M.-X. Wei, C.-A. Chen, H.-P. Lin, C.-H. Hsu, Synthesis of multiporous carbons from the water caltrop shell for high-performance supercapacitors, *ACS Omega*, 5 (2020) 10626–10632.

[4] L. Doherty, Y. Zhao, X. Zhao, Y. Hu, X. Hao, L. Xu, R. Liu, A review of a recently emerged technology: Constructed wetland - Microbial fuel cells, *Water Research*, 85 (2015) 38–45.

[5] Z. Xu, S. Du, Z. Yi, J. Han, C. Lai, Y. Xu, X. Zhou, Water chestnut-derived slope-dominated carbon as a high-performance anode for high-safety potassium-ion batteries, *ACS Applied Energy Materials*, 3 (2020) 11410–11417.

[6] P. Wang, L. Fan, L. Yan, Z. Shi, Low-cost water caltrop shell-derived hard carbons with high initial coulombic efficiency for sodium-ion battery anodes, *Journal of Alloys and Compounds*, 775 (2019) 1028–1035.

[7] C.-C. Hsu, Y.-C. Lin, Y.-Y. Lin, H.-T. Li, C.-S. Ni, C.-I. Liu, C.-C. Chang, L.-C. Lin, Y.-T. Pan, S.-F. Liu, T.-Y. Liu, H.-Y. Chen, Trapa natans husk-derived nanoporous carbons as electrode materials for sustainable high-power microbial fuel cell supercapacitor systems, *Advanced Energy and Sustainability Research*, 3 (2022) 2100163.

[8] F.-Y. Lin, Y.-Y. Lin, H.-T. Li, C.-S. Ni, C.-I. Liu, C.-Y. Guan, C.-C. Chang, C.-P. Yu, W.-S. Chen, T.-Y. Liu, H.-Y. Chen, Trapa natans husk-derived carbon as a sustainable electrode material for plant microbial fuel cells, *Applied Energy*, 325 (2022) 119807.

[9] F. Rodríguez-Reinoso, The role of carbon materials in heterogeneous catalysis, *Carbon*, 36 (1998) 159–175.

[10] M.A. Yahya, Z. Al-Qodah, C.W.Z. Ngah, Agricultural bio-waste materials as potential sustainable precursors used for activated carbon production: A review, *Renewable and Sustainable Energy Reviews*, 46 (2015) 218–235.

[11] W.M.A.W. Daud, W.S.W. Ali, M.Z. Sulaiman, The effects of carbonization temperature on pore development in palm-shell-based activated carbon, *Carbon*, 38 (2000) 1925–1932.

[12] A. Ahmadpour, D.D. Do, The preparation of active carbons from coal by chemical and physical activation, *Carbon*, 34 (1996) 471–479.

[13] H. Marsh, F. Rodríguez-Reinoso, Chapter 2 - Activated carbon (origins), in: H. Marsh, F. Rodríguez-Reinoso (Eds.) *Activated Carbon*, Elsevier Science Ltd, Oxford, 2006, pp. 13–86.

[14] F. Caturla, M. Molina-Sabio, F. Rodríguez-Reinoso, Preparation of activated carbon by chemical activation with $ZnCl_2$, *Carbon*, 29 (1991) 999–1007.

[15] R.-L. Tseng, Mesopore control of high surface area NaOH-activated carbon, *Journal of Colloid and Interface Science*, 303 (2006) 494–502.

[16] T. Budinova, E. Ekinci, F. Yardim, A. Grimm, E. Björnbom, V. Minkova, M. Goranova, Characterization and application of activated carbon produced by H_3PO_4 and water vapor activation, *Fuel Processing Technology*, 87 (2006) 899–905.

[17] O. Oginni, K. Singh, G. Oporto, B. Dawson-Andoh, L. McDonald, E. Sabolsky, Influence of one-step and two-step KOH activation on activated carbon characteristics, *Bioresource Technology Reports*, 7 (2019) 100266.

[18] X. Zhu, F. Qian, Y. Liu, D. Matera, G. Wu, S. Zhang, J. Chen, Controllable synthesis of magnetic carbon composites with high porosity and strong acid resistance from hydrochar for efficient removal of organic pollutants: An overlooked influence, *Carbon*, 99 (2016) 338–347.

[19] I. Major, J.-M. Pin, E. Behazin, A. Rodriguez-Uribe, M. Misra, A. Mohanty, Graphitization of Miscanthus grass biocarbon enhanced by in situ generated FeCo nanoparticles, *Green Chemistry*, 20 (2018) 2269–2278.

[20] Z. Zhao, C. Ma, F. Chen, G. Xu, R. Pang, X. Qian, J. Shao, X. Hu, Water caltrop shell-derived nitrogen-doped porous carbons with high CO_2 adsorption capacity, *Biomass and Bioenergy*, 145 (2021) 105969.

[21] C. Ma, J. Bai, M. Demir, X. Hu, S. Liu, L. Wang, Water chestnut shell-derived N/S--doped porous carbons and their applications in CO_2 adsorption and supercapacitor, *Fuel*, 326 (2022) 125119.

[22] C. Yang, W. Que, Y. Tang, Y. Tian, X. Yin, Nitrogen and sulfur Co-doped 2D titanium carbides for enhanced electrochemical performance, *Journal of the Electrochemical Society*, 164 (2017) A1939.

[23] K.-J. Tsai, C.-S. Ni, H.-Y. Chen, J.-H. Huang, Single-walled carbon nanotubes/Ni–Co–Mn layered double hydroxide nanohybrids as electrode materials for high-performance hybrid energy storage devices, *Journal of Power Sources*, 454 (2020) 227912.

[24] Y.-Y. Tung, S. Gull, C.-S. Ni, W.-J. Chiu, H.-Y. Chen, Recent progress in stretchable and self-healable supercapacitors: active materials, mechanism, and device construction, *Journal of Micromechanics and Microengineering*, 32 (2022) 073001.

[25] M. Armand, J. Tarascon, Issues and challenges facing rechargeable lithium batteries, *Nature*, 414 (2001) 359–367.

[26] C.-C. Lin, H.-Y. Liu, J.-W. Kang, C.-C. Yang, C.-H. Li, H.-Y.T. Chen, S.-C. Huang, C.-S. Ni, Y.-C. Chuang, B.-H. Chen, C.-K. Chang, H.-Y. Chen, In-situ X-ray studies of high-entropy layered oxide cathode for sodium-ion batteries, *Energy Storage Materials*, 51 (2022) 159–171.

[27] Y.-M. Chang, Y.-C. Wen, T.-Y. Chen, C.-C. Lin, S.-C. Huang, C.-S. Ni, A.-Y. Hou, C.-W. Hu, Y.-F. Liao, C.-H. Kuo, S.-F. Liu, W.-W. Wu, L.-J. Li, H.-Y. Chen, Understanding charge storage mechanisms for amorphous $MoSnSe_{1.5}S_{1.5}$ nanoflowers in alkali-ion batteries, *Advanced Energy Materials*, 13 (2023) 2301125.

[28] H.-Y. Chen, N. Bucher, S. Hartung, L. Li, J. Friedl, H.-P. Liou, C.-L. Sun, U. Stimming, M. Srinivasan, A multi-walled carbon nanotube core with graphene oxide nanoribbon shell as anode material for sodium ion batteries, *Advanced Materials Interfaces*, 3 (2016) 1600357.

[29] R.A. Adams, J.-M. Syu, Y. Zhao, C.-T. Lo, A. Varma, V.G. Pol, Binder-free N- and O-rich carbon nanofiber anodes for long cycle life K-ion batteries, *ACS Applied Materials & Interfaces*, 9 (2017) 17872–17881.

[30] W. Weng, J. Xu, C. Lai, Z. Xu, Y. Du, J. Lin, X. Zhou, Uniform yolk–shell Fe7S8@C nanoboxes as a general host material for the efficient storage of alkali metal ions, *Journal of Alloys and Compounds*, 817 (2020) 152732.

[31] J. Li, Y. Li, X. Ma, K. Zhang, J. Hu, C. Yang, M. Liu, A honeycomb-like nitrogen-doped carbon as high-performance anode for potassium-ion batteries, *Chemical Engineering Journal*, 384 (2020) 123328.

[32] J. Chen, B. Yang, H. Hou, H. Li, L. Liu, L. Zhang, X. Yan, Disordered, large interlayer spacing, and oxygen-rich carbon nanosheets for potassium ion hybrid capacitor, *Advanced Energy Materials*, 9 (2019) 1803894.

[33] A.K. Yadav, P. Dash, A. Mohanty, R. Abbassi, B.K. Mishra, Performance assessment of innovative constructed wetland-microbial fuel cell for electricity production and dye removal, *Ecological Engineering*, 47 (2012) 126–131.

8 Unlocking the Potential of Biomass Energy

Nguyen Viet Linh Le and Shou-Heng Liu
National Cheng Kung University, Tainan City, Taiwan

With extremely strong development in many fields, the demand for energy has also increased strongly in last few decades. Traditional fuel sources such as fossil fuels, with their outstanding advantages, although very convenient, also cause significant pressure on the environment and society. Therefore, finding alternative fuel sources to help partly overcome the above problems is extremely necessary. Biomass is a fuel source with great potential and is familiar to humans throughout the development process. However, we have only exploited a very small part of the potential of this fuel source. In recent years, a lot of research has been carried out to optimize the process of converting biomass into energy suitable for many different biomass sources to achieve the requirements for application on an industrial scale.

8.1 INTRODUCTION: BACKGROUND AND DRIVING FORCES

Energy has been one of the key factors accompanying the development of mankind. In almost every field, energy has always acted as a sufficient condition to be able to operate all machines and systems before even thinking about developing them. The manufacturing and transportation industries are no exception. However, with the rapid development, along with the trend of globalization, the traditional fuel sources, which play the role of providing energy for most industries, increasingly reveal their weaknesses. Traditional fossil fuel sources have three great disadvantages: first, the risk of depletion; second, causing environmental pollution; and third, causing energy insecurity in some countries. The risk of fossil fuel depletion is not a major concern at present, as reported reserves worldwide are still very large and will continue to increase as technology develops to allow humans to explore and exploit more natural resources. However, the remaining two weaknesses are a burning issue that is being paid much attention at present. While the sharp increase in energy consumption demand directly causes increasingly serious negative impacts on the environment, Covid-19 and political instability in recent years have made the instability of the fossil fuel sources, especially gasoline and oil, are more volatile than ever. To find and shift to renewable resources is essential and is also an area of research that is receiving a lot of attention today [1]. Among alternative fuels, biomass energy is considered an extremely potential fuel source. Biomass, the main source of raw materials for biofuel production, is available almost everywhere and even underutilized. All sources of microorganisms, plants, or animals are biomass and have potential for biofuel production [2]. Agricultural residues and waste from industry, farms, and

DOI: 10.1201/9781003463009-8

households are the largest sources of biomass. The utilization of these by-products and waste sources not only solves the problem of energy but also solves the environmental problems and help the economic circle to be more complete [3].

The European Union (EU) in 2013 suggested a seventh program, "Good quality of life, taking into account the limitations of our planet". Reducing the impact of consumption on the environment, including minimizing food waste and utilizing biomass sustainably, is one of the most significant objectives of these programs [4]. Demirbas and Dincer [5] predicted the shares of alternative fuels compared to the total automotive fuel consumption. However, the energy transition is very slow and not as planned, especially in the automotive industry. According to the European Automobile Manufacturers' Association (ACEA), alternative fuels only accounted for 1.2%. Despite the added competition of electric and hybrid cars in the market reducing the share of diesel fuel from 90.2% to 86%, diesel fuel still largely dominates this market [6]. This makes the process of improving environmental quality stagnant, despite a lot of improvements and mandatory regulations to help vehicles using diesel fuel have a better emission index. Moreover, the continuous occurrence of political conflicts in a number of member countries of the Organization of the Petroleum Exporting Countries (OPEC) further reveals that the instability in the supply of this fuel is a serious threat to energy security in many countries, even the most developed ones. Therefore, biofuel sources, which have already been proven to be potential and relevant, need to be noticed and given the opportunity to become a more important fuel source in the future. In this context, internal combustion engines are a long way from being completely replaced, especially in aircraft or ships, and the dissemination of renewable biofuels is becoming more and more important.

The process to convert biomass into energy is not really too complicated nor too lofty requirements; however, optimization of the processes is still under research. Potential biomass sources as input materials for the energy production process are extremely diverse, however, that is also why different raw material sources require certain changes in the process. Moreover, the finished product of this process includes all three phases, solid, liquid, and gas, which can be used to provide energy. Therefore, depending on the biomass source and target product, different methods can be applied accordingly. Typically, there exist three primary methods for transforming biomass into energy that encompass mechanical conversion, thermal conversion, and biological conversion [7]. However, these methods can be combined with each other or with several other pre-treatment or post-treatment methods to achieve the greatest process efficiency. Understanding the nature as well as the advantages and disadvantages of these processes helps to optimize the design of energy production lines, contributing to promoting the process of popularizing this safe, environmentally friendly, and potential fuel source.

8.2 BIOMASS FEEDSTOCKS AND CONVERSION METHODS

8.2.1 Generations of Biomass

Biomass is an energy source familiar to humans throughout our development. As there is more awareness about energy use methods as well as social environmental issues, people are increasingly careful in choosing and improving biomass fuel sources for use. Until today, there have been four generations of biofuel, corresponding to different biomass sources. Table 8.1 shows four generations of biofuel, their main biomass sources as well as the advantages and disadvantages of each generation. In the past, the first generation of biofuel was extremely popular because this biomass source was extremely abundant, and the method of use was also extremely simple. All organic materials of plant and animal origin and that are flammable can be considered as potential fuel sources, and the way to use them is simply to combust it. However, people soon realized that the abuse of first-generation biofuel would lead to many negative impacts on the environment and society. Therefore, next generations of biofuel are researched and proposed to help use this energy source in a safer and more sustainable way. While first-generation biofuels have adverse effects on the food supply chain and ecological balance, and some even have a very low energy return on investment (EROI), with some even below one, indicating that the energy output is lower than the energy input required, the next generation of biofuels has significantly reduced these negative effects and improved the EROI coefficient. However, the technology to produce these fuels is not yet complete, especially third and fourth generation biofuels. Therefore, next-generation biofuel production plants on an industrial scale are not yet common. Currently, second-generation biofuel is attracting the most attention because of its feasibility in large-scale production.

TABLE 8.1
Generations of Biomass Sources and Their Characteristics

	First Generation	Second Generation	Third Generation	Fourth Generation
Biomass Feedstocks	Edible biomass (wheat, corn, sugar cane, …)	Non-edible biomass (wood, straw, waste, …)	Algal biomass (macroalgae, microalgae)	Genetically modified biomass
Impact on food security	High impact on food supply chain	No impact	No impact	No impact
Impact on biodiversity	High impact on biodiversity	No impact	No impact	No impact
Impact on environmental	High impact on environmental	Depend on biomass feedstocks	Environmentally friendly but only marine eutrophication is a concern	Environmentally friendly but threat of genetic modification is a concern
Energy return on investment (EROI)	Low	Fair to high depend on biomass feedstocks	High	Very high
Commercialization possibility	Existing commercial production	Existing commercial production	Possible but no commercial product yet	Conception and technology just in early stages

Besides, second generation biofuels, although very dependent on environmental conditions, can take advantage of many different waste sources, even municipal waste, which cannot be avoided in daily life and social activities. Meanwhile, the current production efficiency of third-generation biofuels remains low, making it challenging to alter the energy situation. Similarly, although fourth-generation biofuels offer numerous advantages, genetic structure modification remains a significant concern, necessitating further research to assess the extent of its impact.

Conditions for a fuel source to be considered viable and sustainable include three factors: a significant volume, cyclical and short cycles, and the absence of negative effects on society [8]. Having to meet all three of the below criteria makes many biomass sources, which can easily be converted into biofuel, become unsuitable. A large forest of trees is without a doubt a huge source of biomass and can be converted into biofuel extremely easily and quickly. However, the exploitation of such biomass resources can directly have serious environmental and social impacts, not to mention the fact that forests have a low regeneration capacity that takes a lot of time. The current trend that is receiving a lot of attention from scientists as well as governments is converting waste into energy, which is the second generation of biofuels. While other biomass sources consume a significant amount of resources, for example, first-generation biomass sources require a lot of land and water resources to grow crops, waste is almost a perfect resource if it can be exploited. Waste appears in all human activities; it is more than capable of meeting the requirements of large capacity and circulation. Moreover, utilizing waste does not cause negative impacts on the environment; but on the contrary, it helps solve these problems. According to the US Environmental Protection Agency (EPA) [9], municipal solid waste (MSW) generation in 2018 totaled 292.4 million tons, equivalent to 4.9 pounds per person per day. About 69 million tons of MSW produced were recycled, and 25 million tons were composted. Together, over 94 million tons of MSW were recycled and composted, which works out to a rate of 32.1% for both processes. This means that more than two-thirds of MSW waste is not currently treated and can become a potential input source for energy production. Similarly, the waste from agricultural, forestry, and fishery activities also have a huge capacity and potential to exploit. According to statistics, the current amount of biomass waste accounts for approximately 30% of the total annual biomass production, which equates to roughly 1.3 billion tons [10]. However, converting this biomass source is not easy. The components in waste are often unstable and complex, especially municipal solid waste from daily social human activities, which can cause an uneven quality of finished products. Therefore, more research is needed to develop an efficient and smooth process of turning waste into energy. Chandrasiri et al. [11] suggested producing bioethanol based on waste. Research results show that wastepaper types are very suitable for bioethanol production. In another study, Sohaib et al. [12] proposed an option to convert agricultural waste sources, such as rice straw and corn stover, into biofuel through fast pyrolysis. The study also found the most suitable conditions to create biohydrogen from this agricultural waste. From a different perspective, Wang et al. [13] analyzed the possibilities for using pelletized biomass made from various types of municipal solid waste as fuel. The study used four different types of waste, including dog manure, horse manure, apple pomace waste, and tea waste, combined with an industrial by-product (NovoGro) to create solid fuel pellets. Research results show that this combination is completely feasible when the combined samples have positive results and have the potential to be used as fuel (except apple pomace because the high content of K, Ca, and Si will give a low melting point and produce slag). In addition, because of its large particle sizes and high cellulose and hemicellulose contents, these materials had a high total moisture, low softening temperature and low calorific value. These properties are not suitable to be used as fuel. In general, the use of municipal solid waste as a raw material for biofuel production is not only significant in terms of energy solutions, but also in terms of waste management.

Third generation biofuel is another approach that has attracted a lot of attention from researchers in recent years. These biofuels are typically derived from feedstocks like algae, aquatic plants, and certain microorganisms and are completely unrelated to the global food supply chain. Biomass feedstocks of this type of fuel are also quite diverse and accessible, moreover, the finished product yield is also quite high. Some studies show that, with the same production area, some third-generation biofuel production processes can be 7 to 31 times more efficient than the next best crop—palm oil [14]. Another advantage is that these fuel production processes usually do not cause negative impacts on the environment, one of the prerequisites for achieving sustainable energy goals. However, because it is a new and not yet perfected approach, the production cost of third generation biofuel is often quite expensive and cannot compete with the second generation. Furthermore, the stability of this fuel is also a factor that needs to be considered. Not only the technical parameters of the production process, but also other hard-to-control objective factors such as weather, temperature, humidity, or light can affect process performance [15]. If the process can be improved to reduce the production cost of third-generation biofuel as well as maintain its stability in the future, this type of fuel can play a huge role in the global energy picture.

Fourth generation biofuel is the concept to upgrade third generation biofuel. These fuels result from combining genetically modified microorganisms with genetically engineered source materials [16]. They have all the advantages of third-generation biofuels, plus their yields are greatly improved and they can also be produced in harsh environments. However, genetic modification is a risky approach and needs to be strictly controlled. Scientists fear that improved genetic resources could leak and destroy the biological balance in nature [17]. It can be said that fourth generation biofuels are a potential direction in the future. However, at the present time, they are only stopped at the conceptual level.

8.2.2 Biomass Conversion Methods

For the purpose of converting biomass feedstock into energy, there are three methods including mechanical conversion, thermal conversion, and biological conversion. The finished products of these three processes mostly exist in the form of biochar (solid), bio-oil (liquid) and biogas (gas). However, the main product and the ratio of each phase in the finished product are completely different. Figure 8.1 shows biomass-to-energy conversion methods and their signature products [8]. Among the

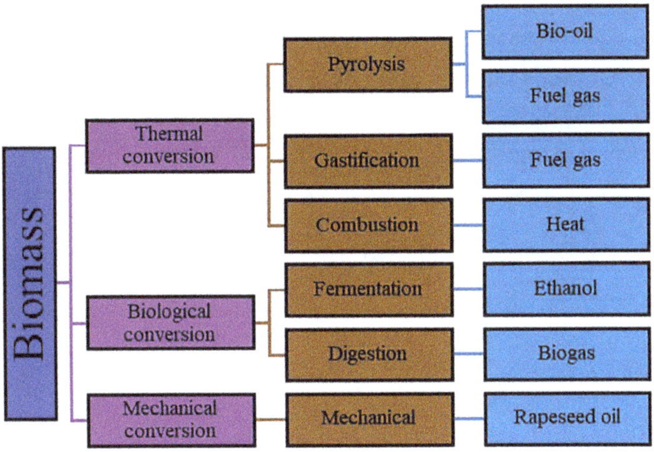

FIGURE 8.1 Biomass conversion methods and their products.

methods listed, gasification and pyrolysis are the most popular and have attracted the most attention due to their technological adaptability, high conversion effects, potentially competitive costs, and significant flexibility in terms of scale of operations and product range. In case manufacturers want to focus on producing specific products such as biogas or ethanol, biological conversion is also a potential choice. Moreover, in order to optimize biofuel production, it is possible to combine two or more different methods. For example, mechanical methods are often used as a method of pretreating biomass so that the following stages can take place smoothly instead of becoming the main method for the whole process of biomass conversion. Mechanical conversion alone can only be applied on a very small or individual scale because of its lack of competitiveness. This section will update the latest research improvements of each method and delve into the analysis of each method to clarify the challenges and opportunities for organizations when they want to exploit biomass and produce biofuel on an industrial scale.

8.2.2.1 Mechanical Conversion

Mechanical conversion is one of the rudimentary methods of converting biomass to biofuel. This method hardly changes the properties of the biomass, so the efficiency is often very low. The main purpose of mechanical conversion is only to reduce the unit size and to maximize the surface area so that it can be easily transported or used. Besides, mechanical methods are an indispensable part to treat biomass before and after applying other biomass conversion methods. Usually, the most popular mechanical methods to reduce the size of biomass are dry milling, compression milling, vibratory ball milling, wet milling, etc. Reducing the size of biomass will significantly increase their total contact surface area, making thermal conversion or biological conversion processes smoother and more thorough [18, 19]. Table 8.2 summarizes some mechanical methods used as size reduction pretreatment for biomass.

TABLE 8.2
Mechanical Methods for Size Reduction Biomass

Biomass Type	Mechanical Methods	Results	Ref.
Grass feed, straw and stalk	Hammer mills	• Higher bulk density of biomass • Higher surface area of lignocellulosic biomass	[22]
Wheat straw	Centrifugal and ball milling	• Higher surface area, lower energy consumption, and particle size	[23]
Wheat and barley straw, corn stoves, etc.	Hammer mill,	• The mill screen size negatively correlated to specific energy consumption. • Higher the content of moisture, larger the specific energy consumption • Greater the screen openings, greater the mean particle diameter of the grinds	[24]
Wheat, maize, and soybean	Multi-cracker system	• Disc speed, disc type, gap between the grinding discs, and type of materials affected the mean particle size	[24]

In addition to the effect of size reduction of biomass, mechanical methods are also used as a mixing method in combination with other biomass treatment methods to maximize conversion efficiency. Wu et al. [20] applied intermittent ball milling combined with enzymatic hydrolysis to effectively convert lignocellulosic biomass into glucose, which is then converted into biofuel through biological fermentation. According to research findings, ball milling accelerated the enzymatic cleavage of cellulosic biomass, resulting in a threefold reduction in processing time. In another study, Du et al. [21] created a horizontal rotating bioreactor for enzymatic hydrolysis of pretreated maize stover to enhance the process economics of ethanol production. This bioreactor released glucose more efficiently with different modes of experiment than the vertical stir-tank reactor. Table 8.3 summarizes several mechanical mixing methods to improve the process of biofuel production.

However, the disadvantage of most mechanical methods is that the implementation cost is often very high, mostly because of the cost of electricity. Therefore, besides conducting further research to improve performance and reduce costs, businesses should consider and calculate carefully when using mechanical methods to pretreat biomass.

TABLE 8.3
Mechanical Mixing Methods to Improve the Process of Biofuel Production

Biomass	Mixing Method	Glucose Yield (%)	Ref.
Aspen wood	Intermittent ball milling	84.7	[20]
Corn stover	Double helical impeller	~6	[21]
Corn stover	Rushton impeller	~40	[25]
Wheat straw	Segmented helical stirrer	76	[26]

8.2.2.2 Thermal Conversion

Significantly more complex than mechanical conversion, thermal conversion involves many processes and chemical reactions. Therefore, they are divided into 3 smaller types depending on temperature and reaction conditions, including combustion, pyrolysis, and gasification.

Combustion is the simplest and oldest way humans use to convert biomass into energy. Burning a piece of wood for heating is one of the most intuitive examples to visualize the combustion process. From an industrial perspective, combustion has also been used for a long time, and almost every country has combustion plants to produce energy. Depending on the combustion system design, they will be divided into three types: grate-firing, pulverized fuel, and fluidized bed [27]. Grate-firing is more popular and is often used on small and medium scales, while pulverized fuel and a fluidized bed are often applied on larger scales. To optimize process efficiency, biomass feedstocks will typically go through many pre-treatment steps, such as grinding and drying before combustion. However, despite efforts to reduce emissions, the conventional combustion process always releases large amounts of exhaust gas and soot. This directly affects the environment and faces a lot of opposition from people, environmental protection organizations as well as governments around the world. In the future, there is a high possibility that traditional combustion plants will be completely eliminated and replaced by other, cleaner, and more sustainable ways of producing energy.

Meanwhile pyrolysis is a more complicated chemical process in which biomass feedstocks are heated in the absence of oxygen. This process breaks down the organic substances into simpler chemical compounds, typically in the form of gases, liquids, and solid residues. The temperature of the pyrolysis process is usually between 300 °C and 800 °C. Because of the absence of oxygen, biomass does not combust or burn, but instead, it thermally decomposes. A series of chemical reactions will occur simultaneously during pyrolysis, including dehydration, isomerization, dehydrogenation, aromatization, charring, oxidation, and other processes. The main product of this process is usually bio-oil. Even under some ideal conditions and high-quality biomass feedstocks, the liquid phase component can account for 60–75% on a weight basis [28]. Therefore, if the desired end product is bio-oil, pyrolysis stands out as one of the primary options.

Pyrolysis is very sensitive to processing input parameters, especially temperature. It is not surprising that temperature plays the most important role in a thermal conversion process. However, increasing temperature does not always improve pyrolysis efficiency. For different target products, the suitable temperature ranges will also change. For example, increasing to too high a temperature will tend to increase the gas phase product, which is detrimental if the desired product is bio-oil. Table 8.4 shows some studies for optimizing liquid will range from 300 °C to 600 °C.

Besides temperature, other important parameters that need special attention include heating rate, residence time, reaction time, particle size, and quality of biomass. Although not as important as temperature, these parameters still directly affect the yield and quality of bio-oil [40]. Optimizing these parameters can help minimize energy input, directly reducing production costs to make bio-oil more competitive and attractive in the energy market. Therefore, although pyrolysis is not a new

TABLE 8.4
Optimal Temperatures for Pyrolysis of Different Biomass

Biomass	Optimal Temperature (°C)	Liquid Yield	Ref.
Mango seed almond	650	38.8%	[29]
Bamboo sawdust	405	72%	[30]
Rice straw	445	68%	[30]
Cotton	510	55%	[31]
Neem de-oiled cake	400	40.2%	[32]
Cynara cardunculus L.	400	56.23%	[33]
Olive bagasse	600	72.4%	[33]
Rice husk	450	70%	[34]
Poplar	455	69%	[35]
Cassava stalk	469	61.39%	[36]
Cassava rhizome	472	63.23%	[36]
Sugarcane bagasse	475	56%	[37]
Jatropha seed shell cake	470	48%	[38]
Residues from palm tree	500	72.4%	[39]

process, research to comprehensively evaluate the effects of the above factors on process performance as well as design optimal setups for each type of biomass is still ongoing and has been intensively researched in recent years.

Gasification is a thermochemical conversion process similar to pyrolysis; however, there are some changes in process parameters. Therefore, the composition of the finished product also changes. One of the biggest differences between gasification and pyrolysis is that pyrolysis is carried out in the absence of oxygen, while on the contrary, gasification is carried out in an environment with the presence of oxygen or steam. Therefore, the main product of the gasification process is syngas, including carbon dioxide, water, carbon monoxide, hydrogen, and gaseous hydrocarbons. The temperature of the gasification process is also higher than pyrolysis, usually around 700–1500 °C. The main reason is because the process requires a larger amount of heat to convert biomass feedstocks into syngas. Gasification is often chosen if the target product is syngas, especially for hydrogen production. Gasification enhances the worth of the raw materials by converting feedstock that is of low value or even negative value into marketable fuels and products [41]. Because there are many similarities with pyrolysis, the product and performance of the gasification process are also affected by similar factors, such as temperature, heating rate, residence time, pressure, etc [42].

From an industrial perspective, the first designs of gasification reactors appeared in the 1920s and were continuously improved to meet larger production needs as well as to minimize negative factors to the environment throughout the process. Figure 8.2 shows the development of a gasifier. In the past, fuel production through gasification has not received as much attention as pyrolysis for two main reasons. First, the energy input requirements of gasification are greater, and second, syngas is also more difficult to transport and store than bio-oil. Both of the above reasons directly increase the

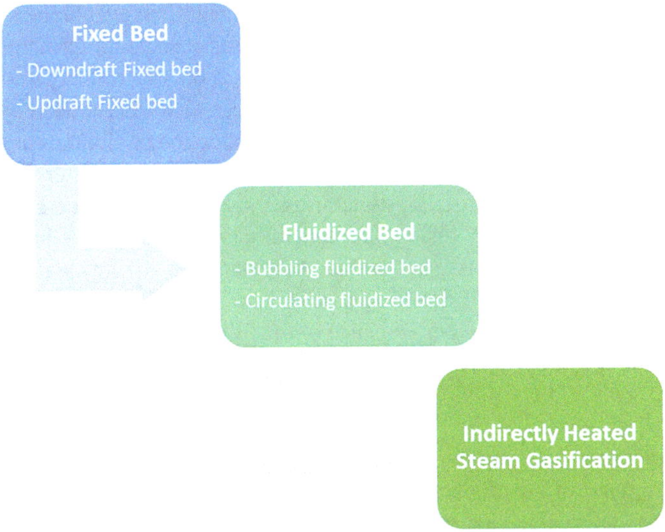

FIGURE 8.2 The development of gasification reactors.

cost of this fuel source, making it less attractive in the eyes of consumers. However, thanks to research efforts to improve process efficiency and the development of technology in recent years, the above problems are gradually being resolved and opening up a possible path to contribute to solving energy problems.

8.2.2.3 Biological Conversion

Biochemical conversion of biomass includes breaking down biomass into gaseous or liquid fuels, such as biogas or bioethanol, using bacteria, microorganisms, and enzymes. The two biochemical processes that are most widely used are fermentation and anaerobic digestion (or biomethanation). Organic material, such as human waste, is broken down during anaerobic digestion through the metabolic pathways of naturally existing microbes in an oxygen-deprived environment. Biological processes can produce a huge variety of compounds as well as clean energy. According to the biorefinery concept, products of biological processes can be found in many forms: biofuels, either liquid (e.g., alcohols and alkanes) or gaseous (e.g., hydrogen and methane), value-added products (e.g., carotenoids, omega-3 and omega-6 fatty acids) acids and antioxidants) and other chemical building blocks (e.g., acetic acid and lactic acid). These compounds can result from processing feedstocks as lignocellulosic or microbial biomass, as well as a wide range of wastes and residues (e.g., forestry and agriculture, industrial processing, municipal solids, sewage, and animal manure) [43].

Fermentation, or more precisely industrial fermentation usually describes the process operations that utilize a chemical change induced by a living organism or enzyme, in particular bacteria, yeast, molds or fungi which produce a specified product. Some of the most important and noticed products of industrial fermentation are alcohols (especially ethanol) and biohydrogen.

Alcohol is an extremely familiar product of fermentation, even appearing a lot in daily life. However, to produce alcohols on an industrial scale by fermentation is not simple. Besides ethanol, the alcoholic fermentation product has the greatest economic benefit, other alcohols such as methanol or butanol can also be produced by fermentation on an industrial scale. Raw materials for the production of alcohols are extremely diverse and widespread, such as agricultural residues (corn stover, crop straws, and bagasse), herbaceous crops (alfalfa, switchgrass), short rotation wood crops, forestry, municipal and agro-industrial residues or generally known as lignocellulosic biomass [43]. Figure 8.3 shows the steps to produce bioethanol by the fermentation process. Lignocellulosic biomass will first be pretreated for screening and characterization, with the aim of improving the performance of the following processes. After pretreatment, the biomass is hydrolyzed by enzymes to convert polysaccharides into monomer sugars such as pentoses (C5 sugars like xylose) and hexoses (C6 sugars like glucose). The products of the final hydrolysis process will be fermented under the action of microorganisms to produce bioethanol—the final product.

Biohydrogen can be produced by biological processes such as dark or photo-fermentation of organic compounds or photolysis of water performed by certain microalgae and cyanobacteria. Nitrogenases and bidirectional hydrogenases are frequently involved in the synthesis of biohydrogen. The interaction of these enzymes is crucial for fermentation-based hydrogen generation. Interestingly, dark fermentation and photo-fermentation, two seemingly opposite methods, can both produce biohydrogen. While the requirement for dark fermentation is the absence of light, the presence of light becomes a mandatory condition for photo-fermentation. Not surprisingly, the bacterial strains required for these two processes are different. While dark fermentation uses mainly Enterobacter and Clostridium bacteria strains [44, 45], photo-fermentation uses green algae such as Chlamydomonas reinhardtii, cyanobacteria such as Anabaena sp., and bacteria from the genus Rhodobacter as Rhodobacter sphaeroides [44, 46]

Fermentation is an easy process where foods, waste, garbage, etc., in daily life still regularly undergo fermentation without any human effort. However, industrial fermentation is extremely complex and needs to be carried out carefully and strictly managed so as not to create unwanted products. In addition, any error in implementation can cause process failure or product quality and quantity to be severely reduced. In addition, the input materials of the fermentation process, if not strictly managed, will also cause negative impacts on the environment.

Meanwhile, digestion or anaerobic digestion (AD) is a process through which bacteria break down organic matter such as animal manure, wastewater biosolids, and food wastes in the absence of oxygen. Special reactors are created in a variety of forms and sizes, depending on the location and feedstock circumstances, and are used to conduct anaerobic digestion for the production of biogas. Complex microbial

FIGURE 8.3 Steps to produce bioethanol by fermentation process.

populations are found in these reactors, which digest waste to create biogas and digestate, the solid and liquid byproducts of the AD process, which are released from the digester. Anaerobic digestion is the least energy-consuming process among all bioenergy production technologies with GHG emission reduction benefit [47]. One of the most important and most studied products of anaerobic digestion is methane. Methane, like many other biofuels, produces few atmospheric pollutants and generates less carbon dioxide per unit energy in comparison to other fossil fuels. Table 8.5 summarizes methane yield from several different microalgae species.

From the data of Table 8.5, it can be seen that, although using the same microalgae species, with different studies, the methane yield has a difference, even a very large difference. This is because the initial conditions have a huge impact on the performance of the process. One of the biggest factors reducing the efficiency of the process is the poor yield of high-quality biomass. Therefore, increasing the biomass production is one method for raising the CH_4 yield per unit of cropland. The biomass output would be increased by choosing biomass species that are suitable for a given region and have higher photosynthetic efficiency, better input utilization (for example, fertilizer and irrigation), and have greater resistance to diseases and pests [59]. Another major challenge for lignocellulosic biomass conversion is the complexity of the lignocellulosic biomass structure. Therefore, biomass should be harvested at the proper time, which would not only provide the good yield of biomass but also provide the biomass which could be converted into CH_4 without requiring extensive

TABLE 8.5
Methane Yield from Different Microalgae Species

Microalgae Species	Methane Yield	Loading Rate	Ref.
Arthrospira platensis	481 mL g^{-1} VS	2000 mg/TS/L	[48]
Chlamydomonas reinhardtii	587 mL g^{-1} VS	2000 mg/TS/L	[48]
Chlorella kessleri	335 mL g^{-1} VS	2000 mg/TS/L	[48]
Arthrospira maxima	173 mL g^{-1} VS	500 mg/TS/L	[49]
Chlorella sorokiniana	212 mL g^{-1} VS	N/R	[50]
Chlorella vulgaris	403 mL g^{-1} VS	2 g/VS/L	[51]
Chlorella vulgaris	286 mL g^{-1} VS	5000 mg/VS/L	[52]
Chlorella vulgaris	240 mL g^{-1} VS	1000 mg/VS/L	[53]
Chlorella vulgaris	189 mL g^{-1} VS	N/R	[50]
Euglena gracilis	485 mL g^{-1} VS	2000 mg/TS/L	[48]
Microcystis sp.	70.33–153.51 ml	1500–6000 mg/VS	[54]
Nannochloropsis salina (lipid extracted biomass)	130 mL g^{-1} VS	2000 mg/l/VS	[55]
Phaeodactylum tricornutun	0.35 L g^{-1} COD	1.3 ± 0.4–5.8 ± 0.9	[56]
Scenedesmus obliquus	287 mL g^{-1} VS	2000 mg/TS/L	[48]
Scenedesmus obliquus	240 mL g^{-1} VS	2000 mg/VS/L	[56]
Scenedesmus sp.	170 mL g^{-1} COD	1000 mg/COD/L	[57]
Scenedesmus sp. (single stage)	290 mL g^{-1} VS	18,000 mg/VS/L	[58]
Scenedesmus sp. (two stage)	354 mL g^{-1} VS	18,000 mg/VS/L	[58]

Note: 46 mL/g/VS Hydrogen.

pretreatment of the feedstock. Another restriction for digesting lignocellulosic biomass is the lack of a competent digester for handling high solids feedstocks like this material. The wastewater treatment plant contributed significantly to the present anaerobic digester designs, which were primarily created for handling low solids feedstocks. Operating a lignocellulosic biomass digester of this kind requires a lot of energy, especially while mixing the biomass, which eventually leads to reduced net energy production. Therefore, the energy balance of lignocellulosic biomass digestion to CH_4 might be improved with a suitable digester design capable of effectively processing high solids feedstocks.

8.3 CONCLUSION

Biomass energy resources are extremely diverse. The way to convert them into biofuel energy forms is also very flexible, depending on the origin and purpose of use. Methods can be combined to improve performance as well as obtain the most target final products. The processes used to process biomass into biofuel need to be invested in for further research and improvement before they can be applied on an industrial level. Therefore, we have only been able to take advantage of a very small part of this resource due to incomplete technologies and unclear support policies. As mentioned, utilizing these energy sources not only ensures the energy security of countries, avoiding dependence on other countries rich in natural resources, but also solves economic, social, and environmental issues at the same time.

REFERENCES

[1] A. T. Hoang *et al.*, "Impacts of COVID-19 pandemic on the global energy system and the shift progress to renewable energy: Opportunities, challenges, and policy implications," *Energy Policy*, vol. 154, p. 112322, Jul. 2021, doi: 10.1016/J.ENPOL.2021.112322

[2] A. T. Hoang *et al.*, "Microbial fuel cells for bioelectricity production from waste as sustainable prospect of future energy sector," *Chemosphere*, vol. 287, p. 132285, Jan. 2022, doi: 10.1016/J.CHEMOSPHERE.2021.132285

[3] A. T. Hoang *et al.*, "Perspective review on municipal solid waste-to-energy route: Characteristics, management strategy, and role in circular economy," *J. Clean. Prod.*, vol. 359, p. 131897, Jul. 2022, doi: 10.1016/J.JCLEPRO.2022.131897

[4] "Potential of biomass-to-fuel conversion technologies for power and means of transport." https://sciendo.com/article/10.5604/01.3001.0012.2815 (accessed Apr. 12, 2023).

[5] A. Demirbas and K. Dincer, "Sustainable green diesel: A futuristic view," vol. 30, no. 13, pp. 1233–1241, Aug. 2008, doi: 10.1080/15567030601082829

[6] "Fuel types of new vans: electric 5.3%, diesel 86.0% market share full-year 2022 - ACEA - European Automobile Manufacturers' Association." https://www.acea.auto/fuel-cv/fuel-types-of-new-vans-electric-5-3-diesel-86-0-market-share-full-year-2022/ (accessed Apr. 12, 2023).

[7] T. Bridgwater, "Biomass for energy," *J. Sci. Food Agric.*, vol. 86, no. 12, pp. 1755–1768, Sep. 2006, doi: 10.1002/JSFA.2605

[8] T. Bich, N. Nguyen, N. Viet, and L. Le, "Biomass resources and thermal conversion biomass to biofuel for cleaner energy: A review," *J. Emerg. Sci. Eng.*, vol. 1, no. 1, pp. 6–13, Sep. 2023, doi: 10.61435/JESE.2023.2

[9] "National overview: Facts and figures on materials, wastes and recycling | US EPA." https://www.epa.gov/facts-and-figures-about-materials-waste-and-recycling/national-overview-facts-and-figures-materials (accessed Nov. 07, 2022).
[10] J. Gustavsson, C. Cederberg, and U. Sonesson, "*Global Food Losses and Food Waste*," 2011.
[11] Y. S. Chandrasiri, W. M. L. I. Weerasinghe, D. A. T. Madusanka, and P. M. Manage, "Waste-based second-generation bioethanol: A solution for future energy crisis," *Int. J. Renew. Energy Dev.*, vol. 11, no. 1, pp. 275–285, Feb. 2022, doi: 10.14710/IJRED.2022.41774
[12] Q. Sohaib, M. Habib, Syed Fawad Ali Shah, U. Habib, and S. Ullah, "Fast pyrolysis of locally available green waste at different residence time and temperatures," vol. 39, no. 15, pp. 1639–1646, Aug. 2017, doi: 10.1080/15567036.2017.1363830
[13] T. Wang *et al.*, "Evaluation of the potential of pelletized biomass from different municipal solid wastes for use as solid fuel," *Waste Manag.*, vol. 74, pp. 260–266, Apr. 2018, doi: 10.1016/J.WASMAN.2017.11.043
[14] A. Alaswad, M. Dassisti, T. Prescott, and A. G. Olabi, "Technologies and developments of third generation biofuel production," *Renew. Sustain. Energy Rev.*, vol. 51, pp. 1446–1460, Nov. 2015, doi: 10.1016/J.RSER.2015.07.058
[15] M. Abbasi, M. S. Pishvaee, and S. Mohseni, "Third-generation biofuel supply chain: A comprehensive review and future research directions," *J. Clean. Prod.*, vol. 323, p. 129100, Nov. 2021, doi: 10.1016/J.JCLEPRO.2021.129100
[16] R. Kumar, A. K. Ghosh, R. Dhurandhar, and S. Chakrabortty, "Downstream process: Toward cost/energy effectiveness," *Handb. Biofuels*, pp. 249–260, Jan. 2022, doi: 10.1016/B978-0-12-822810-4.00012-9
[17] B. Abdullah *et al.*, "Fourth generation biofuel: A review on risks and mitigation strategies," *Renew. Sustain. Energy Rev.*, vol. 107, pp. 37–50, Jun. 2019, doi: 10.1016/J.RSER.2019.02.018
[18] J. Y. Zhu and X. J. Pan, "Woody biomass pretreatment for cellulosic ethanol production: Technology and energy consumption evaluation," *Bioresour. Technol.*, vol. 101, no. 13, pp. 4992–5002, Jul. 2010, doi: 10.1016/J.BIORTECH.2009.11.007
[19] Y. Zheng, Z. Pan, and R. Zhang, "Overview of biomass pretreatment for cellulosic ethanol production," *Int. J. Agric. Biol. Eng.*, vol. 2, no. 3, pp. 51–68, Sep. 2009, doi: 10.25165/IJABE.V2I3.168
[20] Y. Wu *et al.*, "Application of intermittent ball milling to enzymatic hydrolysis for efficient conversion of lignocellulosic biomass into glucose," *Renew. Sustain. Energy Rev.*, vol. 136, p. 110442, Feb. 2021, doi: 10.1016/J.RSER.2020.110442
[21] J. Du *et al.*, "Enzymatic liquefaction and saccharification of pretreated corn stover at high-solids concentrations in a horizontal rotating bioreactor," *Bioprocess Biosyst. Eng.*, vol. 37, no. 2, pp. 173–181, Jun. 2013, doi: 10.1007/S00449-013-0983-6
[22] L. G. Tabil, P. K. Adapa, and M. Kashaninejad, "Biomass feedstock pre-processing–Part 1: Pre-treatment," *Biofuel's Engineering Process Technology*, Dr. Marco Aurelio Dos Santos Bernardes editors. ISBN: 978-953-307-480-1, InTech 2011. 2011.
[23] A. Barakat, S. Chuetor, F. Monlau, A. Solhy, and X. Rouau, "Eco-friendly dry chemo-mechanical pretreatments of lignocellulosic biomass: Impact on energy and yield of the enzymatic hydrolysis," *Appl. Energy*, vol. 113, pp. 97–105, Jan. 2014, doi: 10.1016/J.APENERGY.2013.07.015
[24] S. Mani, L. G. Tabil, and S. Sokhansanj, "Grinding performance and physical properties of wheat and barley straws, corn stover and switchgrass," *Biomass Bioenergy*, vol. 27, no. 4, pp. 339–352, Oct. 2004, doi: 10.1016/J.BIOMBIOE.2004.03.007

[25] J. Zhang, D. Chu, J. Huang, Z. Yu, G. Dai, and J. Bao, "Simultaneous saccharification and ethanol fermentation at high corn stover solids loading in a helical stirring bioreactor," *Biotechnol. Bioeng.*, vol. 105, no. 4, pp. 718–728, Mar. 2010, doi: 10.1002/BIT.22593
[26] D. Ludwig, B. Michael, T. Hirth, S. Rupp, and S. Zibek, "High solids enzymatic hydrolysis of pretreated lignocellulosic materials with a powerful stirrer concept," *Appl. Biochem. Biotechnol.*, vol. 172, no. 3, pp. 1699–1713, Nov. 2013, doi: 10.1007/S12010-013-0607-2
[27] L. Rosendahl, *Biomass Combustion Science, Technology and Engineering.*
[28] A. Demirbas, "Recent advances in biomass conversion technologies," *Energy Edu. Sci. Technol.*, vol. 6, pp. 19–41, 2000.
[29] E. Lazzari *et al.*, "Production and chromatographic characterization of bio-oil from the pyrolysis of mango seed waste," *Ind. Crops Prod.*, vol. 83, pp. 529–536, May 2016, doi: 10.1016/J.INDCROP.2015.12.073
[30] S.-H. Jung, B.-S. Kang, and J.-S. Kim, "Production of bio-oil from rice straw and bamboo sawdust under various reaction conditions in a fast pyrolysis plant equipped with a fluidized bed and a char separation system", doi: 10.1016/j.jaap.2008.04.001
[31] J. Lu Zheng, W. Ming Yi, and N. Na Wang, "Bio-oil production from cotton stalk," *Energy Convers. Manag.*, vol. 49, no. 6, pp. 1724–1730, Jun. 2008, doi: 10.1016/J.ENCONMAN.2007.11.005
[32] V. Volli and R. K. Singh, "Production of bio-oil from de-oiled cakes by thermal pyrolysis," *Fuel*, vol. 96, pp. 579–585, Jun. 2012, doi: 10.1016/J.FUEL.2012.01.016
[33] J. M. Encinar, J. F. González, and J. González, "Fixed-bed pyrolysis of Cynara cardunculus L. Product yields and compositions," *Fuel Process. Technol.*, vol. 68, no. 3, pp. 209–222, Dec. 2000, doi: 10.1016/S0378-3820(00)00125-9
[34] J. Alvarez, G. Lopez, M. Amutio, J. Bilbao, and M. Olazar, "Bio-oil production from rice husk fast pyrolysis in a conical spouted bed reactor," *Fuel*, vol. 128, pp. 162–169, Jul. 2014, doi: 10.1016/J.FUEL.2014.02.074
[35] J. Makibar, A. R. Fernandez-Akarregi, M. Amutio, G. Lopez, and M. Olazar, "Performance of a conical spouted bed pilot plant for bio-oil production by poplar flash pyrolysis," *Fuel Process. Technol.*, vol. 137, pp. 283–289, Sep. 2015, doi: 10.1016/J.FUPROC.2015.03.011
[36] A. Pattiya and S. Suttibak, "Production of bio-oil via fast pyrolysis of agricultural residues from cassava plantations in a fluidised-bed reactor with a hot vapour filtration unit," *J. Anal. Appl. Pyrolysis*, vol. 95, pp. 227–235, May 2012, doi: 10.1016/J.JAAP.2012.02.010
[37] M. R. Islam, M. Parveen, and H. Haniu, "Properties of sugarcane waste-derived bio-oils obtained by fixed-bed fire-tube heating pyrolysis," *Bioresour. Technol.*, vol. 101, no. 11, pp. 4162–4168, Jun. 2010, doi: 10.1016/J.BIORTECH.2009.12.137
[38] S. W. Kim *et al.*, "Bio-oil from the pyrolysis of palm and Jatropha wastes in a fluidized bed," *Fuel Process. Technol.*, vol. 108, pp. 118–124, Apr. 2013, doi: 10.1016/J.FUPROC.2012.05.002
[39] N. Abdullah and H. Gerhauser, "Bio-oil derived from empty fruit bunches," *Fuel*, vol. 87, no. 12, pp. 2606–2613, Sep. 2008, doi: 10.1016/J.FUEL.2008.02.011
[40] A. T. Hoang *et al.*, "Progress on the lignocellulosic biomass pyrolysis for biofuel production toward environmental sustainability," *Fuel Process. Technol.*, vol. 223, p. 106997, Dec. 2021, doi: 10.1016/J.FUPROC.2021.106997
[41] M. Puig-Arnavat, J. C. Bruno, and A. Coronas, "Review and analysis of biomass gasification models," *Renew. Sustain. Energy Rev.*, vol. 14, no. 9, pp. 2841–2851, Dec. 2010, doi: 10.1016/J.RSER.2010.07.030

[42] R. S. El-Emam, I. Dincer, and G. F. Naterer, "Energy and exergy analyses of an integrated SOFC and coal gasification system," *Int. J. Hydrogen Energy*, vol. 37, no. 2, pp. 1689–1697, Jan. 2012, doi: 10.1016/J.IJHYDENE.2011.09.139

[43] L. Gouveia and P. C. Passarinho, "Biomass conversion technologies: Biological/biochemical conversion of biomass," *Lect. Notes Energy*, vol. 57, pp. 99–111, Mar. 2017, doi: 10.1007/978-3-319-48288-0_4/COVER

[44] I. K. Kapdan and F. Kargi, "Bio-hydrogen production from waste materials," *Enzyme Microb. Technol.*, vol. 38, no. 5, pp. 569–582, Mar. 2006, doi: 10.1016/J.ENZMICTEC.2005.09.015

[45] O. Elsharnouby, H. Hafez, G. Nakhla, and M. H. El Naggar, "A critical literature review on biohydrogen production by pure cultures," *Int. J. Hydrogen Energy*, vol. 38, no. 12, pp. 4945–4966, Apr. 2013, doi: 10.1016/J.IJHYDENE.2013.02.032

[46] A. E. Marques, A. T. Barbosa, J. Jotta, M. C. Coelho, P. Tamagnini, and L. Gouveia, "Biohydrogen production by Anabaena sp. PCC 7120 wild-type and mutants under different conditions: Light, nickel, propane, carbon dioxide and nitrogen," *Biomass Bioenergy*, vol. 35, no. 10, pp. 4426–4434, Oct. 2011, doi: 10.1016/J.BIOMBIOE.2011.08.014

[47] I. Hamawand, "Anaerobic digestion process and bio-energy in meat industry: A review and a potential," *Renew. Sustain. Energy Rev.*, vol. 44, pp. 37–51, Apr. 2015, doi: 10.1016/J.RSER.2014.12.009

[48] J. H. Mussgnug, V. Klassen, A. Schlüter, and O. Kruse, "Microalgae as substrates for fermentative biogas production in a combined biorefinery concept," *J. Biotechnol.*, vol. 150, no. 1, pp. 51–56, Oct. 2010, doi: 10.1016/J.JBIOTEC.2010.07.030

[49] A. E. Inglesby and A. C. Fisher, "Enhanced methane yields from anaerobic digestion of Arthrospira maxima biomass in an advanced flow-through reactor with an integrated recirculation loop microbial fuel cell," *Energy Environ. Sci.*, vol. 5, no. 7, pp. 7996–8006, Jun. 2012, doi: 10.1039/C2EE21659K

[50] G. Polakovičová, P. Kušnír, S. Nagyová, and J. Mikulec, "Process integration of algae production and anaerobic digestion," in *15th International Conference on Process Integration, Modelling and*, 2012, vol. 29.

[51] F. Lü, J. Ji, L. Shao, and P. He, "Bacterial bioaugmentation for improving methane and hydrogen production from microalgae," *Biotechnol. Biofuels*, vol. 6, no. 1, pp. 1–11, Jul. 2013, doi: 10.1186/1754-6834-6-92/FIGURES/6

[52] A. M. Lakaniemi, C. J. Hulatt, D. N. Thomas, O. H. Tuovinen, and J. A. Puhakka, "Biogenic hydrogen and methane production from Chlorella vulgaris and Dunaliella tertiolecta biomass," *Biotechnol. Biofuels*, vol. 4, no. 1, pp. 1–12, Sep. 2011, doi: 10.1186/1754-6834-4-34/FIGURES/5

[53] M. Ras, L. Lardon, S. Bruno, N. Bernet, and J. P. Steyer, "Experimental study on a coupled process of production and anaerobic digestion of Chlorella vulgaris," *Bioresour. Technol.*, vol. 102, no. 1, pp. 200–206, Jan. 2011, doi: 10.1016/J.BIORTECH.2010.06.146

[54] S. Zeng, X. Yuan, X. Shi, and Y. Qiu, "Effect of inoculum/substrate ratio on methane yield and orthophosphate release from anaerobic digestion of Microcystis spp.," *J. Hazard. Mater.*, vol. 178, no. 1–3, pp. 89–93, Jun. 2010, doi: 10.1016/J.JHAZMAT.2010.01.047

[55] S. Park and Y. Li, "Evaluation of methane production and macronutrient degradation in the anaerobic co-digestion of algae biomass residue and lipid waste," *Bioresour. Technol.*, vol. 111, pp. 42–48, May 2012, doi: 10.1016/J.BIORTECH.2012.01.160

[56] C. Zamalloa, J. De Vrieze, N. Boon, and W. Verstraete, "Anaerobic digestibility of marine microalgae Phaeodactylum tricornutum in a lab-scale anaerobic membrane bioreactor," *Appl. Microbiol. Biotechnol.*, vol. 93, no. 2, pp. 859–869, Oct. 2011, doi: 10.1007/S00253-011-3624-5

[57] C. González-Fernández, B. Sialve, N. Bernet, and J. P. Steyer, "Thermal pretreatment to improve methane production of Scenedesmus biomass," *Biomass Bioenergy*, vol. 40, pp. 105–111, May 2012, doi: 10.1016/J.BIOMBIOE.2012.02.008

[58] Z. Yang, R. Guo, X. Xu, X. Fan, and S. Luo, "Hydrogen and methane production from lipid-extracted microalgal biomass residues," *Int. J. Hydrogen Energy*, vol. 36, no. 5, pp. 3465–3470, Mar. 2011, doi: 10.1016/J.IJHYDENE.2010.12.018

[59] R. E. H. Sims, A. Hastings, B. Schlamadinger, G. Taylor, and P. Smith, "Energy crops: current status and future prospects," *Glob. Chang. Biol.*, vol. 12, no. 11, pp. 2054–2076, Nov. 2006, doi: 10.1111/J.1365-2486.2006.01163.X

9 Challenges and Solutions in Hydrogen Energy

Nguyen Viet Linh Le and Shou-Heng Liu
National Cheng Kung University, Tainan City, Taiwan

Hydrogen is one of the most promising renewable energy sources to replace traditional fossil fuel. In addition to the highest energy density by mass, the molecular structure of hydrogen is extremely simple and does not contain components that can create harmful emissions, such as carbon or sulfur. However, one of the huge barriers in large-scale applications as fuel source is that the production technologies are still not perfect. Currently, most hydrogen production processes still cause certain impacts on the environment, especially grey hydrogen production processes. To achieve each goal of sustainable development, including minimizing environmental risks, scientists have proposed many solutions to produce green hydrogen, which is discussed in this chapter.

9.1 INTRODUCTION: BACKGROUND AND DRIVING FORCES

Renewable energy in recent years has become one of the sizzling topics when the problems of air pollution from traditional fuel consumption could not be much improved. Also, the unstable supply due to political instability is becoming more and more evident. The dominance of the internal combustion engine in production and transportation activities makes the position of fossil fuel even more solid. However, the increasingly aggressive policies of governments around the world in an effort to ameliorate environmental issues such as greenhouse gas (GHG), global warming, or air pollution have become the driving force for potential alternative energy. Among them, hydrogen, a fuel source that does not contain carbon and can be used directly in internal combustion engines without requiring engine modification, is one of the extremely promising fuel sources. Depending on the production procedure and the degree of impact on the environment, hydrogen is usually classified into 3 types, i.e., grey, blue, and green hydrogen. Table 9.1 compares the difference between grey, blue, and green hydrogen. Grey hydrogen is produced from fossil fuel through steam methane reforming (SMR). The production of grey hydrogen releases a large amount of carbon dioxide, which is one of the major components of greenhouse gas emissions, causing negative environmental impacts. Similarly, blue hydrogen is also produced through the SMR process. However, carbon dioxide is captured and managed in the case of blue hydrogen, greatly reducing the risk of air pollution during production. Yet, because fossil fuel is not a renewable fuel source, grey hydrogen is also not

DOI: 10.1201/9781003463009-9

TABLE 9.1
Comparisons between Various Source of Hydrogen

	Grey Hydrogen	Blue Hydrogen	Green Hydrogen
Raw materials	Fossil fuel (methane)	Fossil fuel (methane)	Any renewable resources like wind, solar or biomass
Method	Steam methane reforming (SMR)	Steam methane reforming (SMR) with carbon capture and storage (CCS)	Water splitting, pyrolysis, gasification, photobiological, photoelectrochemical
CO_2 emission	Emit a large amount of CO_2 (8 kg of CO_2 per kg of H_2)	95–97% CO_2 captured or reused	Potential for net-zero emission
Overall efficiency	70–80%	70–80%	35–50%
Cost	~1.3$/kg	1.6–2$/kg	2–7.5$/kg

considered a renewable fuel source. Unfortunately, more than 95% of all hydrogen produced worldwide depends on fossil fuel, and grey and blue hydrogen account for half of that [1]. As a result, the energy problem cannot be improved by using grey hydrogen. In the opposite direction, green hydrogen is generated from renewable energy sources such as solar energy, wind energy, or biomass resources which are considered extremely eco-friendly. As such, they not only have the potential to achieve the goal of carbon neutrality but also solve other serious energy problems. However, the approach is still emerging and the development of facilities as well as technology has not kept pace with energy demand. In addition, the issue of reinvestment costs becomes an issue upon industrializing this source of fuel. Because there are so many different approaches, green hydrogen has a very wide range of production costs depending on the raw materials used, ranging from $2 to $7.50 to produce one kilogram of hydrogen [2–7]. Meanwhile, it costs only about 1.3$ to produce the same amount of grey hydrogen and from 1.6 to 2$ in the case of blue hydrogen [8]. In addition, the overall yield for green hydrogen production is only 35–50% depending on the production method, which is significantly lower than that of grey and blue hydrogen when the efficiency of the SMR process is at least 70%. However, with advances in both production technology and improved energy conversion efficiency, green hydrogen is expected to be economically competitive with other hydrogen sources in the near future [9]. Within the scope of this chapter, the production and application aspects of green hydrogen are discussed, since it is the only hydrogen source that can help create a sustainable and circular economy and achieve the net-zero emission goal.

9.2 GREEN HYDROGEN PRODUCTION

Hydrogen is the simplest and lightest element, also a colorless, odorless, tasteless substance but extremely flammable. It is commonly seen on Earth, especially in the composition of water and organic compounds. With the desire to maximize production efficiency, the research also focuses on the hydrogen conversion process from these

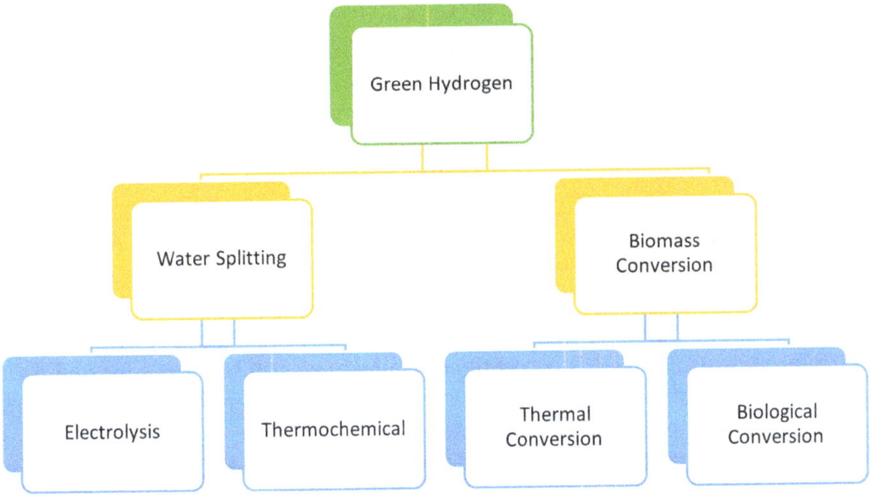

FIGURE 9.1 Different approaches to produce green hydrogen.

raw materials. Therefore, the two most common approaches in hydrogen production are (1) splitting water into hydrogen and oxygen by applying electrolysis or thermochemical and (2) converting biomass to hydrogen through thermal conversion or biological conversion, as shown in Figure 9.1. This section will introduce and discuss the advantages and disadvantages of different green hydrogen production methods.

9.2.1 Hydrogen from Biomass Conversion

Hydrogen produced from biomass is an approach that not only solves energy problems but also helps to solve social problems. Biomass is a general term used to refer to any organic matter derived from the earth's living organisms. Therefore, biomass sources used for raw materials are very diverse, from first generation with edible sources such as wheat, corn, sugar cane crops to second generation with non-edible sources such as wood, grass, or by-product and waste product from agriculture, forestry, and fishing activities. Moreover, even municipal solid waste (MSW) is considered as one of the viable biomass sources even though their composition is extremely complex and unstable. With the third generation, algae can be used as a feedstock for energy production, with the several advantages. For example, it can be cultivated on non-arable land or in wastewater to minimize the competition for land and resources. However, currently the second-generation biomass together with the thermal conversion method is getting more attention by researchers because of its high efficiency and much shorter production time when compared to the biochemical conversion of third-generation biomass.

9.2.1.1 Thermal Conversion

To produce hydrogen from biomass via thermal conversion, the two most efficient methods are pyrolysis and gasification. Although these two methods have some similarities, there are notable differences between pyrolysis and gasification. While pyrolysis occurs in the absence or with limited oxygen, gasification must be performed under conditions of oxygen or steam. Therefore, the product distributions are also different. The main products of pyrolysis are bio-oil, syngas, and biochar, whereas gasification is primarily focused on the production of syngas.

Pyrolysis is an environmentally sensitive process, i.e., any small change can have a significant impact on process performance. Understanding the influence of different operating conditions will help to optimize the selection of target products. Temperature is usually the most crucial factor for pyrolysis because it supplies the energy needed to break the chemical interactions between the various biomass components. Based on operating conditions, pyrolysis is also divided into different types. Figure 9.2 shows the different kinds of pyrolysis to produce hydrogen.

Conventional pyrolysis is a traditional method and has been widely used for a long time. Depending on the heating rate (HR), conventional pyrolysis is divided into 4 different types, including slow, intermediate, fast, and flash pyrolysis. The heating rate will directly affect the temperature in the reactor, the residence time of the substances, and thus also change the phase distribution of the products. Table 9.2 shows the operation parameters of slow, intermediate, fast, and flash pyrolysis. Conventional

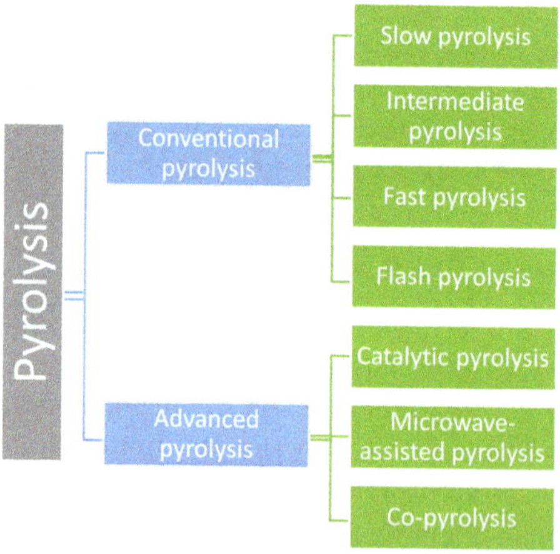

FIGURE 9.2 Different kinds of biomass pyrolysis to produce hydrogen.

TABLE 9.2
Operation Parameters of Different Pyrolysis

	Operating Conditions		
	Temperature (°C)	**Residence Time (s)**	**Heating Rate (°C/s)**
Slow pyrolysis	300–700	>300	<1
Intermediate pyrolysis	450–550	2–10	<100
Fast pyrolysis	450–800	0.5–10	10–200
Flash pyrolysis	800–1200	<0.5	>1000

pyrolysis is an effective method to convert biomass into biofuel because of its short reaction time, no complex equipment and infrastructure requirements, and simple steps. However, for the purpose of hydrogen production, conventional pyrolysis has many limitations. First, the concentration of gas phase products (in particular, hydrogen) in the conventional pyrolysis is not high. The main product of the process is usually the liquid phase, which often accounts for 40–50% or more. The distribution of the solid phase in the product composition is also very high, often accounting for more than 20%. The gas phase (especially the hydrogen yield) is considered only by-products in the process and is not usually focused on. Furthermore, due to the fact that pyrolysis is an endothermic process and biomass generally has a limited ability to transfer heat, conventional pyrolysis requires a substantial input of heat. However, the efficiency of this process is often compromised as the heat primarily concentrates on the surface and struggles to effectively reach the inner regions. This is a major challenge to design an efficient pyrolysis system for hydrogen production at an industrial scale. Therefore, many different options have been proposed to improve pyrolysis as well as enhance the selectivity toward hydrogen production.

One of the most popular ways to improve performance of the chemical process is using catalysts. In the presence of a catalyst, not only is the activation energy of the reaction reduced but also the reaction rate is enhanced. Besides, the catalyst also helps to change the selectivity of the reaction, which can simultaneously reduce the formation of undesirable products from the solid phase and promote the formation and increase of hydrogen yields. There are various kinds of catalysts that can be used in the biomass pyrolysis process, such as zeolite, alkali and alkaline earth metals, noble metal, metal oxide, etc. Each kind of catalyst has its own advantages and disadvantages, as shown in Table 9.3.

Another advanced method of pyrolysis is microwave-assisted pyrolysis (MAP). Microwave-assisted pyrolysis is one of the most effective and promising methods as it

TABLE 9.3
Advantages and Disadvantages of Different Catalysts

Category	Catalyst	Advantages	Disadvantages
Zeolite	ZSM-5; HZSM-5	High catalytic activity; selectivity control and stability	Catalyst deactivation and costly
Alkali and Alkaline earth metals	K, Mg, Ca, Na	High catalytic activity; high selectivity control; rapid reaction rates; lower reaction temperatures	Catalyst deactivation; limited applicability to specific feedstocks and environmental concerns
Metal catalysts	W, Mo, Fe, Ni, Pd, Ru, Au, Rh	High catalytic activity; various metal option; catalytic versatility and high resistance with catalyst poisoning	Limited selectivity control; environmental concerns and some noble metals are very expensive
Metal oxide catalysts	Al_2O_3, ZrO_2, TiO_2, CeO_2, CaO	High thermal stability; catalytic versatility and high resistance with catalyst poisoning	Low surface area; limited selectivity control; catalyst deactivation and environmental concerns

solves one of the biggest disadvantages of conventional pyrolysis. Usually, the energy efficiency of conventional pyrolysis is very poor because the heat will transfer from the surface of biomass particles to the depths of their core and the amount of heat as well as the rate of heat transfer is completely dependent on the thermal conductivity of the biomass. However, one of the common characteristics of biomass is its thermal conductivity only at an average level, which makes most of the wasted heat radiated to the environment instead of transferring into the biomass. Therefore, the pyrolysis process requires a large amount of energy to complete. In addition, the great quantity of solid phase products is observed, which is often not targeted and has very low value. However, with the microwave-assisted pyrolysis approach, the heat penetrates the biomass and generates heat within it more efficiently. Moreover, microwave energy is directly absorbed by the reactants, reducing the energy losses associated with heating bulk materials or heating up the reactor itself. Figure 9.3 shows the difference of heat transfer between microwave-assisted pyrolysis and conventional pyrolysis.

Co-pyrolysis involves the combination of biomaterials with different physicochemical properties and subjecting them to pyrolysis together. Similar to other pyrolysis processes, some catalysts can be added to the process to improve the reaction efficiency. The advantage of co-pyrolysis is that biomass types can offset each other's shortcomings to improve product yields and properties. In addition, the co-pyrolysis process can be performed simply, which may not require any improvement over the conventional pyrolysis process [10]. However, to find proper combinations with acceptable performance, their effective hydrogen index (EHI) needs to be calculated and considered. The EHI is a parameter that shows the ratio of hydrogen to carbon in a compound, considering the elimination of heteroatoms to produce NH_3, H_2S, and H_2O. This relationship can be mathematically represented by the following equation [11]:

$$EHI = \frac{H - 2O - 3N - 2S}{C} \tag{9.1}$$

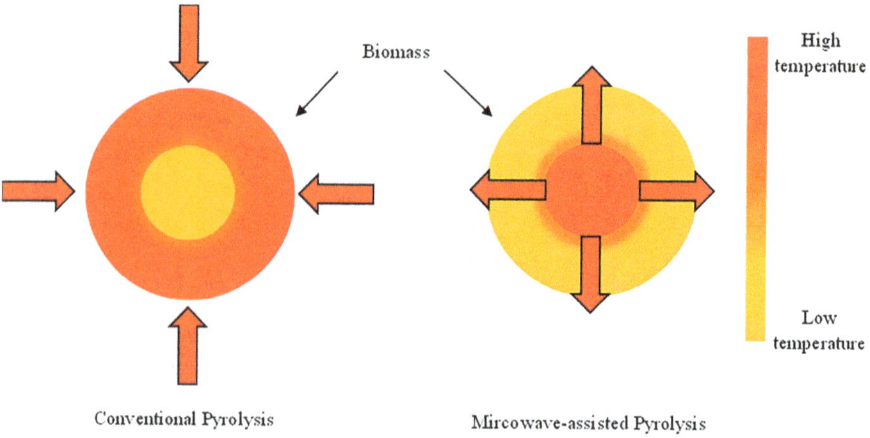

FIGURE 9.3 Difference of heat transfer between microwave-assisted pyrolysis and conventional pyrolysis.

where C, O, H, N, and S is the number of moles of carbon oxygen, hydrogen, nitrogen, and sulfur in the biomass feedstock, respectively. The biomass with the EHI index < 0.3 is usually not suitable to participate in the co-pyrolysis process and may cause negative effects to the hydrogen production yield [12]. Co-pyrolysis is believed to be a suitable approach for large scale or industrial scale pyrolysis, as it could diversify the choice of input materials. Nonetheless, co-pyrolysis requires careful selection and compatibility of feedstocks. Different feedstocks have varying thermal characteristics, reactivity, and chemical compositions. Achieving optimal co-pyrolysis conditions for different feedstocks can be challenging and may require extensive investigation and process optimization. Any combination that is not suitable can not only greatly affect the hydrogen production efficiency, but also create many undesirable products.

In contrast to pyrolysis where the reaction occurs in the absence of oxygen, gasification involves transforming biomass into a flammable gas blend by subjecting it to partial oxidation at elevated temperatures, usually ranging from 800 to 900 °C. The gasifying agents are usually oxygen, air, or steam. Recently, supercritical water, which used as a gasifying agent, has also been noticed by researchers because of its superior characteristics. At supercritical conditions (temperature >374 °C, pressure >22 MPa), water exhibits both liquid-like and gas-like properties. It has a high density, low viscosity, and a higher solubility for various organic compounds. These properties enhance the gasification reaction, i.e., making the process highly efficient and resulting in a high concentration of hydrogen. However, supercritical water gasification requires high energy input due to the harsh operating conditions involved in reaching and maintaining the supercritical state of water. Therefore, unlike conventional gasification which has already been industrialized, this technology is still under research. Table 9.4 shows the advantages and disadvantages of different gasifying agents.

Gasification increases the effectiveness of converting solid and liquid phases into gas phases. Without any modifications or alterations to enhance process efficiency, such as catalyst utilization or biomass treatment techniques, gasification typically results in a higher hydrogen yield compared to pyrolysis. However, gasification usually requires more energy input because the reaction usually takes place at higher temperature. Moreover, besides an increase in hydrogen yield, the simultaneous rise in the evolution of other gas phase products like CO, CO_2, and CH_4 can occur during gasification. This results in a lower concentration of hydrogen in the gasification process, necessitating the implementation of methods to manage and address the presence of these undesired products. To address this problem, gasification processes often utilize catalysts, such as alkaline earth metallic catalysts, metal-based catalysts (e.g., Ni, Ce, La), and mineral catalysts.

To have a better comparison between two thermal biomass conversion methods for hydrogen production, Table 9.5 shows that by improving the experimental settings, combined with the use of a number of suitable catalysts, the hydrogen production efficiency from different biomass sources is significantly improved, surpassing much more than simply using pyrolysis or gasification methods. Currently, the volume or moles of hydrogen collected often accounts for more than 60% of the gas phase, specially, Stonor et al. [17] announced that a yield of 92.5% mol can be

TABLE 9.4
Advantages and Disadvantages of Different Gasifying Agents

Gasifying Agents	Advantages	Disadvantages
Oxygen	• Higher efficiency and syngas quality • Impurities reduction	• High cost • Limited oxygen availability • Carbon capture challenges • Corrosion and material compatibility
Air	• Most widely used technology • Cost-effective • Simple operation • Easier carbon capture	• Low efficiency syngas quality • High impurities
Steam	• Higher concentration of hydrogen (H_2) due to the water-gas shift reaction • Low tar formation • High energy efficiency • Flexible feedstocks	• Steam supply • High carbon dioxide formation • Equipment corrosion • Slow reaction and specific reaction kinetics requirement
Supercritical water	• High efficiency • Low tar formation • Hydrogen-rich syngas	• Harsh operating conditions • Corrosion and material compatibility • High energy input • Process complexity • Limited feedstock flexibility

achieved by applying alkaline thermal treatment. Because of its extremely outstanding performance and available biomass resources without conflicting with other social benefits such as the environment or affecting the food supply chain, pyrolysis and gasification are currently the methods for hydrogen production that attract the most attention and potential by the researchers.

9.2.1.2 Biological Conversion

Biochemical conversion of biomass for hydrogen production involves the use of biological processes to convert organic materials (biomass) into hydrogen gas (H_2). This is often achieved through fermentation or biophotolysis. This approach is an excellent way to utilize agricultural residues, food waste, and other organic waste materials, reducing their environmental impact. The process of biological conversion typically results in lower carbon emissions, enabling the production of hydrogen that is both renewable and carbon-neutral [20]. In addition, biochemical conversion also uses third generation biomass, which can be generated under much simpler conditions than first and second-generation biomass, rendering this energy source feasible in most countries. However, unlike thermal conversion, biological conversion requires further refinement and optimization to reach an industrial scale because the efficiency of this process is not high, and the hydrogen production time is considerably longer when compared to the thermal conversion method. While thermal conversion takes only a few minutes to a few hours, biological conversion takes days, weeks, or even longer. It is challenging to reduce this time to be comparable to thermal conversion,

TABLE 9.5
Comparisons between Pyrolysis and Gasification

Biomass Feedstock	Catalyst	Method	Gas Content When Achieve Highest H_2 Yield				Unit	Ref.
			H_2	CH_4	CO_2	CO		
Water hyacinth	Ni/sepiolite catalyst	ex-situ/2 stage pyrolysis	65%	4%	21%	10%	%mol	[13]
Herb residues (Salvia miltiorrhiza and a small amount of Panax notoginseng)	10 wt % Ni/CaO	moving bed pyrolyzer	75%	14%	2%	9%	%vol	[14]
Rice husk	Akaline Thermal Treatment (NAOH)	pyrolysis	80%	18%	1%	1%	%vol	[15]
Wheat straw grass	Akaline Thermal Treatment (NAOH)	pyrolysis	86%	6%	4%	-	%mol	[16]
Cellulose powder	Akaline Thermal Treatment($CA(OH)_2$)+Ni/ZrO_2)	pyrolysis	92.50%	1.80%	1.30%	4.20%	%mol	[17]
Corn stalk	CaO	gasification	61.23%	12.11%	9.13%	15.55%	%vol	[18]
Rice straw	CaO	gasification	60.28%	12.63%	8.75%	16.56%	%vol	[18]
Wheat straw	CaO	gasification	58.69%	13.28%	9.58%	16.34%	%vol	[18]
Peanut shell	CaO	gasification	60.84%	10.58%	11.18%	15.48%	%vol	[18]
Rice husk	CeO_2-Ni-CaO	gasification	85%	4%	3.50%	7.50%	%vol	[19]

due to the inherently longer duration of biochemical processes compared to chemical reactions. However, this issue can be partly resolved when manufacturing on an industrial scale by adopting a cyclic or continuous production process.

Fermentation is one of the biological conversion methods used for biohydrogen production and is usually divided into 2 types, i.e., dark fermentation and photofermentation, depending on the reaction conditions. In dark fermentation, certain microorganisms, typically bacteria, break down complex organic compounds present in biomass to produce hydrogen gas as a metabolic product in the absence of light. Among different kinds of biological conversions, this method is regarded as the most well-developed and promising approach [21]. Although the hydrogen production efficiency of dark fermentation is generally lower than that of other hydrogen production methods, the net energy ratio of this process is still very competitive due to the lower energy input requirements. For comparison, while the energy efficiency of dark fermentation, biocatalyzed electrolysis, and steam methane reforming is 9.6%, 25.7%, and 64%, respectively, their net energy ratios are 1.9, 1.8, and 0.64, respectively [22].

Photofermentation, also known as photosynthetic fermentation or biophotolysis, occurs in the presence of light. Certain photosynthetic microorganisms, such as purple non-sulfur (PNS) bacteria, use light energy to convert organic compounds into hydrogen gas and other byproducts. Photofermentation usually occurs by utilizing solar energy, thereby taking advantage of this free and unlimited energy source. Theoretically, photofermentation can produce high hydrogen yields. However, relying too heavily on light sources can reduce efficiency or complicate the operation setup. Normally, for this process to occur smoothly, the light source needs to have wavelengths of 522 nm, 805 nm, and 850 nm [23]. Recently, a combination of dark fermentation and photofermentation has been proposed to overcome the weaknesses of each method [24, 25]. With this approach, the metabolite products from dark fermentation are inoculated with suitable bacteria for use in the subsequent photofermentation process. The two-stage fermentation design not only potentially increases the overall hydrogen production yield compared to individual dark fermentation or photofermentation processes but also can provide continuous operation without relying too much on light.

Biophotolysis of biomass for hydrogen production involves the use of photosynthetic microorganisms, such as certain types of algae and cyanobacteria, to harness light energy and convert organic biomass into H_2 and O_2 through the process of photosynthesis. Biophotolysis is divided into two main types: direct and indirect. In the process of direct biophotolysis, light energy is directly used to split water into H_2 and O_2 during photosynthesis through following equation:

$$2H_2O + \text{light energy} \rightarrow 2H_2 + O_2 \tag{9.2}$$

Meanwhile, the indirect biophotolysis is a process in which light energy is used during photosynthesis to generate energy-rich compounds, such as carbohydrates or organic acids, rather than directly producing H_2 from water. These energy-rich compounds can then serve as substrates for subsequent hydrogen production processes, such as dark fermentation or photofermentation, where H_2 gas is produced as a metabolic byproduct. The overall pathway for indirect biophotolysis involves two following steps:

$$6H_2O + 6CO_2 + \text{light energy} \rightarrow C_6H_{12}O_6 + 6O_2 \quad (9.3)$$

$$C_6H_{12}O_6 + 6H_2O + \text{light energy} \rightarrow 12H_2 + 6CO_2 \quad (9.4)$$

The hydrogen production efficiency of indirect biophotolysis is much higher than that of direct biophotolysis. In particular, indirect biophotolysis is projected to achieve a 10% efficiency in converting light energy to hydrogen energy under full solar irradiance, whereas direct biophotolysis exhibits less than 1.5% efficiency under identical conditions [26]. However, the efficiency of both processes is still considered insufficient for large-scale hydrogen production, and further research is needed to optimize the processes and increase efficiency.

9.2.2 Water Splitting

Water splitting is another clean and renewable method to produce hydrogen. The water is the feedstock and does not emit harmful greenhouse gases during hydrogen generation. There are two main methods of water splitting for hydrogen production, including electrolysis and photolysis (or photocatalysis). During the electrolysis process, water is subjected to an electric current to facilitate the water-splitting reaction. This occurs within an electrolysis cell, which consists of two electrodes (an anode and a cathode) separated by an electrolyte. When a direct current is applied to the electrodes, hydrogen gas is formed at the cathode through hydrogen evolution reaction (HER), while oxygen gas is produced at the anode through oxygen evolution reaction (OER). However, because of the high overpotential, the reaction kinetic of HER and OER are quite slow [27]. Therefore, different catalysts such as single-atom catalysts (SAC), metal catalysts, metal oxides/hydroxides catalysts, metal phosphides, nitrides, borides, carbides, or metal-free catalysts can be used to improve the reaction in both cathode and anode [28].

Photolysis water splitting, also known as photocatalytic water splitting or photocatalysis, is a chemical process that utilizes a semiconductor material as a photocatalyst to absorb light energy and facilitate the water-splitting reaction. When the photocatalyst is illuminated with light, electron-hole pairs are generated, and these charge carriers drive the water-splitting reaction. Similar to electrolysis water splitting, photolysis water splitting also has a low reaction efficiency and requires the participation of catalysts to improve the process. Normally, TiO_2 catalyst is very popular and widely used in the photolysis water splitting process because the energy level of TiO_2 is suitable to trigger the water splitting reaction with low cost and stability [29, 30]. Nevertheless, the efficiency of photocatalytic water splitting under solar energy remains relatively low because TiO_2 (bandgap = 3.2 eV) can only activate the photocatalyst when exposed to UV light, which constitutes a small fraction of the solar spectrum [31]. Therefore, research efforts to strengthen or find new catalysts have been continuously made in the recent years. In general, finding a cost-effective and scalable source of catalysts is crucial to facilitate the practical use of both electrolysis and photolysis water splitting in industries and still requires more comprehensive evaluation studies.

9.3 THE ROLE OF HYDROGEN IN THE ENERGY SECTOR

As mentioned, hydrogen is a clean, renewable, and high energy density source of energy. This implies that hydrogen energy can play an important role in the future, with a vision toward sustainable and environmentally friendly development. In particular, unlike many other renewable energy sources, hydrogen is one of the few energy sources that can be used directly on internal combustion engines, which is still a large and important part of manufacturing and transportation activities. As a carbon-free fuel source, the combustion of hydrogen in the engine produces absolutely no carbon-based pollutants such as carbon monoxide (CO), carbon dioxide (CO_2) or hydrocarbons (HC), in the exhaust gas. The process of burning hydrogen in the air produces only NO_x, an emission that is difficult to avoid when burning any fuel because of the available composition of nitrogen and oxygen in the air. However, because of its small molecular size and high flammability, hydrogen can leak through many solid materials and cause explosions when exposed to air [32]. Therefore, if using hydrogen directly in an internal combustion engine, it is recommended to be more suitable if applied to public transport vehicles or industrial-scale production machines, which will be able to ensure safety under regular inspection and supervision, instead of use on private vehicles.

With a broader view, hydrogen can act as an end product for many processes, completing the renewable energy cycle. This increases efficiency, reduces production costs, and makes widespread adoption of renewable fuels more feasible. To illustrate, through the implementation of thermal conversion and biological conversion methods, hydrogen is the final product of processing various types of waste, ranging from agricultural, forestry, and fishery waste to municipal waste. This helps organizations have more revenue from waste treatments, indirectly reducing production costs to be able to compete with brown hydrogen. Likewise, hydrogen from the water splitting process can be the end product of wastewater treatment. Many waste-to-hydrogen models have been proposed and proven to be both technologically and economically feasible [33, 34]. More interestingly, hydrogen could be considered as a possible switch for hard-to-control renewable energy sources such as solar and wind energy. These energy sources are often used to produce electricity, but one of the biggest weaknesses of these energy sources is that they are very difficult to control. The difference between daily or seasonal supply and demand causes large amounts of electricity to be wasted because, unlike other fuels, electrical energy is difficult or expensive to store. With a variety of hydrogen production methods, these unlimited and free energy sources can be used more optimally, especially, the amount of electricity produced in excess of the demand can completely provide energy for the hydrogen production process by the electrolysis water splitting method. With the increasingly stringent environmental policies of governments around the world, green hydrogen is gradually becoming possible and promises to change the energy landscape dominated by fossil fuels. In July 2023, Sinopec company launched the world's largest solar-to-hydrogen project based on the electrolysis water splitting method. This can confirm the potential of hydrogen and encourage the improved research in this area.

9.4 CONCLUSIONS

With the increasing demand for clean energy, hydrogen is considered as a potential alternative energy source. However, green hydrogen production currently accounts for less than 5% of total hydrogen production worldwide. This is an opportunity as well as a challenge for efforts to improve green hydrogen production processes to solve current energy problems. Currently, hydrogen production via thermal biomass conversion is considered the most efficient and feasible for industrial scale production. Other approaches, such as biological biomass conversion or water splitting with low production efficiency, need to be improved to be competitive with fossil fuels as well as grey hydrogen.

REFERENCES

[1] S. Atilhan, S. Park, M. M. El-Halwagi, M. Atilhan, M. Moore, and R. B. Nielsen, "Green hydrogen as an alternative fuel for the shipping industry," *Curr. Opin. Chem. Eng.*, vol. 31, p. 100668, Mar. 2021, doi: 10.1016/J.COCHE.2020.100668

[2] B. Olateju and A. Kumar, "Hydrogen production from wind energy in Western Canada for upgrading bitumen from oil sands," 2011, doi: 10.1016/j.energy.2011.09.045

[3] B. Olateju, A. Kumar, and M. Secanell, "A techno-economic assessment of large scale wind-hydrogen production with energy storage in Western Canada," 2016, doi: 10.1016/j.ijhydene.2016.03.177

[4] "Hydrogen on the path to net-zero emissions", Accessed: Jun. 07, 2023. [Online]. Available: https://about.jstor.org/terms

[5] P.-M. Heuser, D. Severin Ryberg, T. Grube, M. Robinius, and D. Stolten, "Techno-economic analysis of a potential energy trading link between Patagonia and Japan based on CO_2 free hydrogen," 2019, doi: 10.1016/j.ijhydene.2018.12.156

[6] H. Blanco, W. Nijs, J. Ruf, and A. Faaij, "Potential for hydrogen and power-to-liquid in a low-carbon EU energy system using cost optimization," 2018, doi: 10.1016/j.apenergy.2018.09.216

[7] D. Milani, A. Kiani, and R. Mcnaughton, "Renewable-powered hydrogen economy from Australia's perspective," 2020, doi: 10.1016/j.ijhydene.2020.06.041

[8] B. Olateju and A. Kumar, "Techno-economic assessment of hydrogen production from underground coal gasification (UCG) in Western Canada with carbon capture and sequestration (CCS) for upgrading bitumen from oil sands. Development of a techno-economic model for UCG-CCS and SMR-CCS. Estimation of H_2 production costs with and without CCS for UCG and SMR. UCG is more economical for H_2 production with CCS. SMR is more cost efficient for H_2 production without CCS," 2013, doi: 10.1016/j.apenergy.2013.05.014

[9] M. Noussan, P. P. Raimondi, R. Scita, M. Hafner, F. Eni, and E. Mattei, "The role of green and blue hydrogen in the energy transition - A technological and geopolitical perspective," 2020, doi: 10.3390/su13010298

[10] A. K. Vuppaladadiyam, H. Liu, M. Zhao, A. F. Soomro, M. Z. Memon, and V. Dupont, "Thermogravimetric and kinetic analysis to discern synergy during the co-pyrolysis of microalgae and swine manure digestate," *Biotechnol. Biofuels*, vol. 12, no. 1, Jun. 2019, doi: 10.1186/S13068-019-1488-6

[11] Q. Xie *et al.*, "Fast microwave-assisted catalytic co-pyrolysis of microalgae and scum for bio-oil production," *Fuel*, vol. 160, pp. 577–582, Nov. 2015, doi: 10.1016/J.FUEL.2015.08.020

[12] A. K. Vuppaladadiyam *et al.*, "Biomass pyrolysis: A review on recent advancements and green hydrogen production," *Bioresour. Technol.*, vol. 364, p. 128087, Nov. 2022, doi: 10.1016/J.BIORTECH.2022.128087

[13] S. Liu, J. Zhu, M. Chen, W. Xin, Z. Yang, and L. Kong, "Hydrogen production via catalytic pyrolysis of biomass in a two-stage fixed bed reactor system," *Int. J. Hydrogen Energy*, vol. 39, no. 25, pp. 13128–13135, Aug. 2014, doi: 10.1016/J.IJHYDENE.2014.06.158

[14] B. Zhao *et al.*, "Catalytic pyrolysis of herb residues for the preparation of hydrogen-rich gas," *Energy Fuels*, vol. 34, no. 2, pp. 1131–1136, Feb. 2020, doi: 10.1021/ACS.ENERGYFUELS.9B02177/ASSET/IMAGES/LARGE/EF9B02177_0005.JPEG

[15] H. Zhou and A. H. A. Park, "Bio-energy with carbon capture and storage via alkaline thermal treatment: Production of high purity H_2 from wet wheat straw grass with CO_2 capture," *Appl. Energy*, vol. 264, p. 114675, Apr. 2020, doi: 10.1016/J.APENERGY.2020.114675

[16] P. Qi *et al.*, "An innovative strategy on co-production of porous carbon and high purity hydrogen by alkaline thermal treatment of rice husk," *Int. J. Hydrogen Energy*, vol. 47, no. 55, pp. 23151–23164, Jun. 2022, doi: 10.1016/J.IJHYDENE.2022.04.257

[17] M. R. Stonor, N. Ouassil, J. G. Chen, and A. H. A. Park, "Investigation of the role of $Ca(OH)_2$ in the catalytic alkaline thermal treatment of cellulose to produce H_2 with integrated carbon capture," *J. Energy Chem.*, vol. 26, no. 5, pp. 984–1000, Sep. 2017, doi: 10.1016/J.JECHEM.2017.07.013

[18] B. Li, H. Yang, L. Wei, J. Shao, X. Wang, and H. Chen, "Hydrogen production from agricultural biomass wastes gasification in a fluidized bed with calcium oxide enhancing," *Int. J. Hydrogen Energy*, vol. 42, no. 8, pp. 4832–4839, Feb. 2017, doi: 10.1016/J.IJHYDENE.2017.01.138

[19] X. Zeng *et al.*, "Hydrogen-rich gas production by catalytic steam gasification of rice husk using CeO_2-modified Ni-CaO sorption bifunctional catalysts," *Chem. Eng. J.*, vol. 441, p. 136023, Aug. 2022, doi: 10.1016/J.CEJ.2022.136023

[20] M. Aziz, A. Darmawan, and F. B. Juangsa, "Hydrogen production from biomasses and wastes: A technological review," *Int. J. Hydrogen Energy*, vol. 46, no. 68, pp. 33756–33781, Oct. 2021, doi: 10.1016/J.IJHYDENE.2021.07.189

[21] R. Łukajtis *et al.*, "Hydrogen production from biomass using dark fermentation," 2018, doi: 10.1016/j.rser.2018.04.043

[22] S. Manish and R. Banerjee, "Comparison of biohydrogen production processes," *Int. J. Hydrogen Energy*, vol. 33, no. 1, pp. 279–286, Jan. 2008, doi: 10.1016/J.IJHYDENE.2007.07.026

[23] I. Akkerman, M. Janssen, J. Rocha, and R. H. Wijjels, "Photobiological hydrogen production: Photochemical eeciency and bioreactor design," *Int. J. Hydrogen Energy*, vol. 27, pp. 1195–1208, 2002, Accessed: Jul. 26, 2023. [Online]. Available: www.elsevier.com/locate/ijhydene

[24] R. Grabarczyk, K. Urbaniec, J. Wernik, and M. Trafczynski, "Evaluation of the two-stage fermentative hydrogen production from sugar beet molasses," *Energies*, vol. 12, no. 21, p. 4090, Oct. 2019, doi: 10.3390/EN12214090

[25] J. Hu, W. Cao, and L. Guo, "Directly convert lignocellulosic biomass to H_2 without pretreatment and added cellulase by two-stage fermentation in semi-continuous modes," *Renew. Energy*, vol. 170, pp. 866–874, Jun. 2021, doi: 10.1016/J.RENENE.2021.02.062
[26] W. Wukovits and W. Schnitzhofer, "Fuels – Hydrogen production|Biomass: Fermentation," *Encycl. Electrochem. Power Sources*, pp. 268–275, Jan. 2009, doi: 10.1016/B978-044452745-5.00312-9
[27] N. T. Suen, S. F. Hung, Q. Quan, N. Zhang, Y. J. Xu, and H. M. Chen, "Electrocatalysis for the oxygen evolution reaction: Recent development and future perspectives," *Chem. Soc. Rev.*, vol. 46, no. 2, pp. 337–365, Jan. 2017, doi: 10.1039/C6CS00328A
[28] Z. Chen, W. Wei, and B. J. Ni, "Cost-effective catalysts for renewable hydrogen production via electrochemical water splitting: Recent advances," *Curr. Opin. Green Sustain. Chem.*, vol. 27, p. 100398, Feb. 2021, doi: 10.1016/J.COGSC.2020.100398
[29] N. Fajrina and M. Tahir, "A critical review in strategies to improve photocatalytic water splitting towards hydrogen production," *Int. J. Hydrogen Energy*, vol. 44, no. 2, pp. 540–577, Jan. 2019, doi: 10.1016/J.IJHYDENE.2018.10.200
[30] A. Kudo and Y. Miseki, "Heterogeneous photocatalyst materials for water splitting," *Chem. Soc. Rev.*, vol. 38, no. 1, pp. 253–278, Dec. 2008, doi: 10.1039/B800489G
[31] C. H. Liao, C. W. Huang, and J. C. S. Wu, "Hydrogen production from semiconductor-based photocatalysis via water splitting," *Catal. 2012, Vol. 2, Pages 490-516*, vol. 2, no. 4, pp. 490–516, Oct. 2012, doi: 10.3390/CATAL2040490
[32] A. Kovač, M. Paranos, and D. Marciuš, "Hydrogen in energy transition: A review," *Int. J. Hydrogen Energy*, vol. 46, no. 16, pp. 10016–10035, Mar. 2021, doi: 10.1016/J.IJHYDENE.2020.11.256
[33] J. Lui, W. H. Chen, D. C. W. Tsang, and S. You, "A critical review on the principles, applications, and challenges of waste-to-hydrogen technologies," *Renew. Sustain. Energy Rev.*, vol. 134, p. 110365, Dec. 2020, doi: 10.1016/J.RSER.2020.110365
[34] S. C. Wijayasekera, K. Hewage, O. Siddiqui, P. Hettiaratchi, and R. Sadiq, "Waste-to-hydrogen technologies: A critical review of techno-economic and socio-environmental sustainability," *Int. J. Hydrogen Energy*, vol. 47, no. 9, pp. 5842–5870, Jan. 2022, doi: 10.1016/J.IJHYDENE.2021.11.226

10 Revolutionizing Energy Sustainability

Unleashing the Potential of rSOC Technology

Shu-Yi Tsai and Kuan-Zong Fung
National Cheng Kung University, Tainan City, Taiwan

10.1 INTRODUCTION

In a world where carbon emissions are tightly regulated, the development of infrastructure capable of storing, processing, and converting variable renewable power into affordable and dependable energy products and services poses a significant and pressing challenge.

Nowadays, there is a pressing need to develop large-scale energy storage and conversion technologies to enhance the efficiency of energy storage and conversion processes and address the mismatch between energy supply and demand. Against the backdrop of energy crises and climate change emerging as major global concerns, the role of these technologies in promoting sustainable development in the global energy economy is crucial.

Hydrogen is of paramount importance in numerous chemical industries, including petroleum refining, ammonia production, oil sands processing, and more. It also serves as a clean fuel for transportation, contributes to the production of nitrogen fertilizers, plays a vital role in semiconductor manufacturing and pharmaceutical development, and finds applications in the aerospace industry, among others. Hydrogen's clean-burning properties make it an attractive alternative fuel for transportation, particularly in fuel cell vehicles. It offers zero-emission mobility, as the only byproduct of hydrogen combustion is water vapor. Moreover, hydrogen fuel cells have the potential to revolutionize the aerospace industry by providing efficient and environmentally friendly power for aircraft. Its utilization in chemical industries, agriculture, semiconductor manufacturing, pharmaceuticals, aerospace, and other sectors underscores its versatility and significance.

Despite significant engineering advances in water-splitting technologies over the past few decades, the currently available commercialized methods for producing green hydrogen still fall short of meeting market demands. This necessitates the development of more robust and cost-effective systems. Other challenges include handling and storage complications, safety issues, transportation difficulties, and dependence on fossil fuels. To reduce reliance on fossil fuels for grey hydrogen production and

DOI: 10.1201/9781003463009-10

align with the priorities of the Paris Agreement, steam electrolysis is widely regarded as a viable technique for large-scale applications in the medium term. Therefore, this review aims to critically discuss the application of high-temperature steam electrolysis as a potential clean energy technology and highlight the main challenges associated with its commercialization, such as production costs, electrochemical performance, energy conversion efficiency, and more.

10.2 HYDROGEN PRODUCTION

Hydrogen can be produced through various methods, including from fossil fuels, biomass, and electrochemical processes [1]. Nowadays, the primary method of hydrogen production is through reforming fossil fuels like natural gas and oil, which is also known as grey hydrogen [1, 2]. The main advantage of this method is its relatively low cost and mature technology, making it widely used in industrial and energy fields. However, this process generates carbon dioxide as a byproduct, contributing to the 'greenhouse effect'.

The production process for blue hydrogen is similar to that of grey hydrogen, involving the reforming of fossil fuels to extract hydrogen. However, in the case of blue hydrogen, carbon capture and storage technology are employed to further reduce carbon emissions [1, 3]. This means that most of the carbon dioxide produced during hydrogen production is captured and stored underground to prevent it from entering the atmosphere, thereby reducing its negative impact on climate change.

Blue hydrogen is considered a transitional hydrogen production method, as it performs better than grey hydrogen in terms of reducing carbon emissions. However, carbon capture and storage technology still face challenges such as cost and technical feasibility. Therefore, research and development of more environmentally friendly and sustainable hydrogen production methods, such as green hydrogen produced through electrolysis of water using renewable energy sources, are also being pursued to achieve a lower carbon energy transition.

Considering these issues, global society is gradually shifting towards more sustainable energy sources, such as renewable energy. Green hydrogen is hydrogen produced using electricity from renewable energy sources. Renewable energy sources include solar, wind, hydro, and biomass energy, which are almost unlimited and have minimal environmental impact. Green hydrogen is made when water (H_2O) is split into hydrogen (H_2) and oxygen (O_2). Water splitting is also known as electrolysis, and requires an energy input.

10.3 CARBON-FREE HYDROGEN PRODUCTION

Closer to real sustainability is green hydrogen. Electrolysis offers a promising pathway for carbon-free hydrogen production using renewable and nuclear resources. The technique employs electrolysis—the separation of hydrogen and oxygen molecules by applying electrical energy to water. This chemical reaction occurs within a device known as an electrolyzer. Electrolyzers come in various sizes, ranging from small appliances that are ideal for decentralized hydrogen production on a small scale, to large-scale facilities that can be directly integrated with renewable or other non-greenhouse-gas-emitting electricity generation methods. Three major methods currently under consideration for electrolytic hydrogen production are alkaline electrolysis, proton exchange membrane (PEM) electrolysis, and ceramic oxide electrolysis.

10.3.1 Alkaline Electrolysis

An alkaline electrolysis cell usually contains two electrodes, an aqueous alkaline as shown in Figure 10.1. Usually an alkaline medium is employed (25~30% KOH) [4–6]. The OH^- produced at cathode are transported to anode side and the H^+ remaining at cathode combine with electrons to form hydrogen gas. The electrolytic reactions that occur on each electrode are given by the following equations:

$$\text{Cathode electrode: } 4H_2O + 4e^- \rightarrow 2H_2 + 4OH^- \tag{10.1}$$

$$\text{Anode electrode : } 4OH^- \rightarrow 2H_2O + 4e^- + O_2 \tag{10.2}$$

$$\text{Total reaction : } H_2O \rightarrow H_2 + \frac{1}{2}O_2 \tag{10.3}$$

From the overall reaction, assuming the current efficiency to be 100%, 2 F of electricity are required to produce 1 mol of H_2 gas, i.e., 0.1268 cm^3/As of H_2 gas at ambient conditions. According to Pierre Millet et al. [7], alkaline electrolysis cells typically have a cell voltage ranging from 1.8 to 2.2 V and operate within the temperature range of 343 to 363 K. This indicates a voltage efficiency of only 68% to 80%.

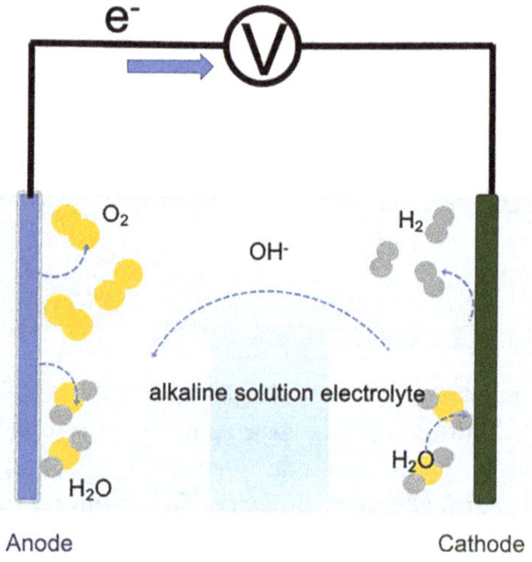

FIGURE 10.1 The scheme and operation of an alkaline electrolysis cell.

10.3.2 Proton Exchange Membrane (PEM)

The adoption of PEM electrolysis started in the 1960s [8], making it a relatively recent technology compared to the Alkaline electrolysis cell. However, this technology has already reached an industrial scale and can achieve nominal outputs exceeding 10MW. The most commonly used membranes are Perfluorosulfonic acid polymer membranes such as Nafion®, Fumapem®, Flemion®, and Aciplex®. However, Nafion® membranes (Nafion® 115, 117, and 212) are predominantly utilized in PEM water electrolyzer due to their significant advantages, such as high current density operation (2A/cm^2), excellent durability, superior proton conductivity, and robust mechanical stability. In a PEM electrolyzer, a Nafion membrane is commonly used as a proton-conducting medium to facilitate the separation of hydrogen and oxygen at the two electrodes as shown in Figure 10.2. At the anode, water is split into protons and oxygen. The generated protons then migrate through the Nafion membrane to the cathode side. At the cathode, the protons combine with electrons to produce hydrogen gas. PEM water electrolysis is accrued by pumping of water to the anode, where it is spilt into oxygen (O_2), protons (H^+) and electrons (e^-). The Reactions are as:

$$\text{Cathode electrode}: 2H^+ + 2e^- \rightarrow H_2 \tag{10.4}$$

$$\text{Anode electrode}: H_2O \rightarrow 2H^+ + \frac{1}{2}O_2 + 2e^- \tag{10.5}$$

$$\text{Total reaction}: 2H_2O \rightarrow H_2 + \frac{1}{2}O_2 \tag{10.6}$$

The PEM electrolyzer was aimed at addressing the operational challenges faced by the alkaline electrolyzer, such as partial load, low current density, and low-pressure

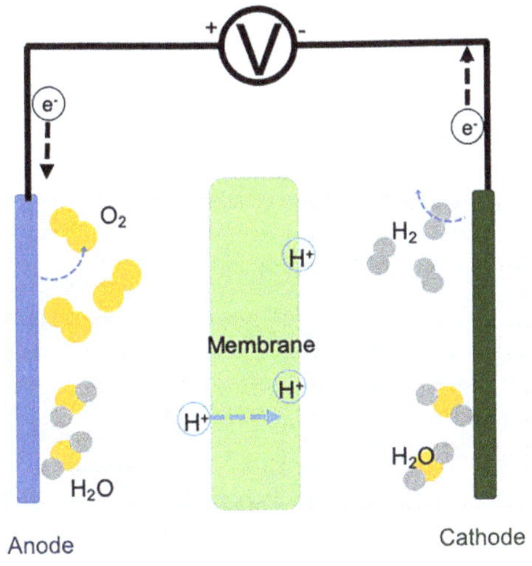

FIGURE 10.2 Schematic illustration of PEM water electrolysis.

operation [9]. In PEM water electrolysis, noble metal-based electrocatalysts, such as Pt/Pd-based catalysts for the hydrogen evolution reaction (HER)[10] at the cathode and RuO_2/IrO_2 catalysts for the oxygen evolution reaction (OER) at the anode [11–13], are commonly utilized [14]. However, the adoption of these expensive materials has contributed to the high cost of PEM electrolysis, despite its drawbacks.

10.3.3 Ceramic Oxide Electrolysis

The introduction of Ceramic oxide electrolysis technology can be traced back to the 1980s, when it was first proposed by Donitz and Erdle [15, 16]. Solid oxide electrolysis (SOE) is a technology that utilizes solid oxide electrolytes to carry out hydrogen production through electrolysis of water. One of the main advantages of SOE technology is its high efficiency. Due to the high conductivity and redox activity of the solid oxide electrolyte, SOE systems can achieve efficient water decomposition at high temperatures, resulting in high conversion efficiency. This makes SOE a promising hydrogen production technology that can play an important role in renewable energy systems. Furthermore, SOE technology also exhibits good flexibility and adjustability. As SOE systems operate at high temperatures (typically between 800°C and 1000°C), they can be adapted to different heat sources, including solar energy, nuclear energy, and industrial waste heat. This enables SOE systems to operate flexibly in different application scenarios and exhibit high adaptability.

10.4 REVERSIBLE SOLID OXIDE CELL

Reversible solid oxide cell (rSOC) technology has emerged as a potential solution for integrating renewable power into the grid, while also offering a pathway for the synthesis of carbon-neutral fuels with high energy density. rSOC devices are electrochemical systems that operate in two distinct and mutually exclusive modes: one for fuel production through chemical inputs and electricity (electrolysis), and another for power generation by consuming fuels (fuel cell mode). The unique advantage of rSOC is its ability to control and mitigate daily electricity demand fluctuations. When the demand for electricity is high, rSOC operates in the Solid Oxide Fuel Cell (SOFC) mode, converting the chemical energy in the fuel directly into electrical energy, thereby reducing the reliance on electrical energy consumption. Conversely, during periods of low electricity demand when power generation systems (such as nuclear power plants, wind power, or solar power systems) have excess power output, rSOC can operate in the Solid Oxide Electrolysis Cell (SOEC) mode. In this mode, electrical and thermal energy are used for high-temperature electrolysis of water or carbon dioxide to produce fuel energy, such as clean hydrogen or syngas. This fuel energy can then be stored and utilized by the fuel cell during peak electricity consumption, ensuring a more efficient and balanced energy supply.

A typical rSOC possess the capability to function as fuel cells, producing electricity from a chemical fuel, or as an electrolyzer, utilizing an electric current to drive an energetically unfavorable chemical reaction (such as the splitting of H_2O into H_2 and O_2). Traditionally, these two operating modes (fuel cell and electrolysis) have been considered as distinct processes necessitating dedicated separate units. However, in the case of solid oxide cells, reversing the direction of the current enables the same device to seamlessly transition between operating as a fuel cell and operating as an electrolyzer. A rSOC cell's ability to switch between its two modes of operation is dependent on the system and cell's capability to supply all reactants to the electrode reaction sites in both modes. There is no inherent limitation in electrochemistry that would prevent a fuel cell from operating in a reversible manner. However, certain implementations are more conducive to nearly reversible operation, while others may encounter difficulties when operating in both directions, due to their design. Typically, they consist of three components arranged in a sandwich-like structure: an air electrode (cathode), an electrolyte, and a fuel electrode (anode) as shown in Figure 10.3.

The electrodes consist of porous layers that facilitate the diffusion of reactants within their structure and catalyze electrochemical reactions. In single technologies such as SOFCs and SOECs, the electrodes have distinct functions, and therefore, they are referred to by their specific names. The anode is the site of the oxidation reaction, while the cathode is where the reduction reaction takes place. On the contrary, reversible solid oxide cells allow for both modes to alternate within the same device. Hence, it is more preferable to use the generic terms "fuel electrode" and "oxygen electrode" instead.

The air electrode and fuel electrode possess a porous structure, which facilitates gas transport and increases the reaction surface area. On the other hand, the electrolyte is completely dense and serves to isolate the air electrode from the fuel electrode, preventing direct gas contact and combustion between the two sides of the electrodes.

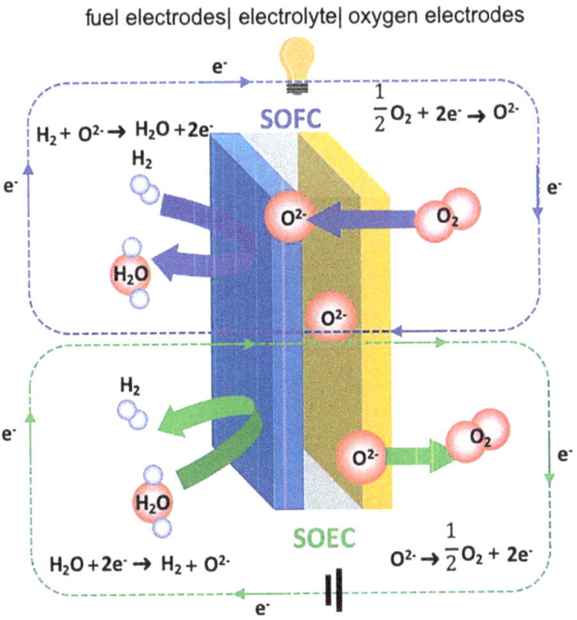

FIGURE 10.3 Configuration of a reversible solid oxide cell.

10.4.1 Oxygen-Ion-Conducting SOECs

According to the ion types conducted by the electrolytes, SOECs can be divided into oxygen-ion-conducting SOECs (O-SOECs) and proton-conducting SOECs (H-SOECs).

Oxygen-ion-conducting solid oxide electrolysis cells (SOECs) are a highly efficient and sustainable technology for hydrogen production. Typically, SOEs consist of an air electrode that produces oxygen at an electric potential of approximately 1.3 V, a fuel electrode that generates hydrogen. For O-SOECs, steam is fed in at the hydrogen electrode side to generate hydrogen; oxygen is produced at the oxygen electrode side as shown in Figure 10.4.

The chemical reactions in an Oxygen-ion-conducting SOECs:

$$\text{Air electrode}: O^{2-} \rightarrow \frac{1}{2}O_2 + 2e^- \quad (10.7)$$

$$\text{Fuel electrode}: H_2O + 2e^- \rightarrow H_2 + O^{2-} \quad (10.8)$$

$$\text{Total reaction}: H_2O \rightarrow H_2 + \frac{1}{2}O_2 \quad (10.9)$$

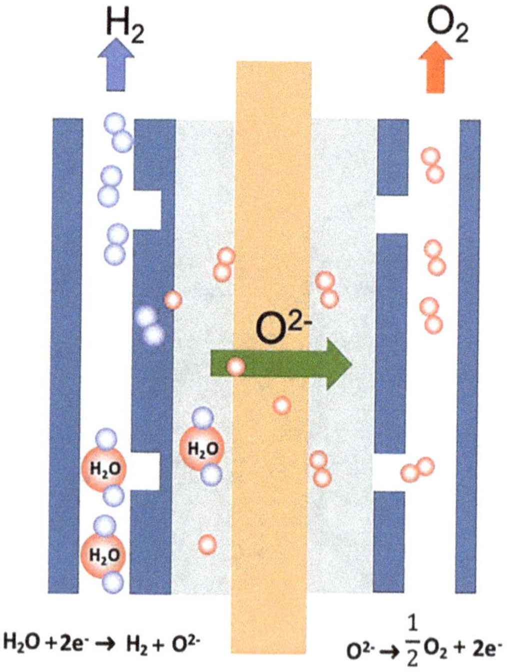

FIGURE 10.4 The basic operation of an oxygen-ion-conducting SOEC with pure hydrogen as a fuel source.

Using external electricity, the H_2O molecule splits to H^+ and O^{2-}, and H^+ combines with e^- to form H_2. The oxygen ions produced are transported through the electrolyte, which is an oxygen ionic conductor. At the anode, the O^{2-} transfers its charge to the anode, and combines with another O^{2-} to form O_2.

Typical O-SOECs consist of an oxygen-ion-conducting solid electrolyte such as yttria-stabilized zirconia (YSZ) sandwiched between two electrodes; fuel electrode made typically of nickel + YSZ and oxygen electrode made typically of perovskite, such as Sr-doped $LaMnO_3$ (LSM) mixed with YSZ. Oxygen-ion-conducting solid oxide electrolysis cells present numerous advantages compared to other electrolysis cell types. Primarily, they function at elevated temperatures ranging from 500 to 900°C. Consequently, these oxygen-ion-conducting SOECs can attain enhanced conversion efficiencies.

Moreover, oxygen-ion-conducting SOECs exhibit excellent stability and durability. The use of oxygen-ion-conducting electrolytes ensures good chemical stability and resistance to degradation, even under harsh operating conditions. This extends the lifespan of the cells and reduces the need for frequent maintenance or replacement.

Another advantage of oxygen-ion-conducting SOECs is their compatibility with renewable energy sources. These cells can efficiently utilize excess electricity generated from renewable sources, such as solar or wind power, to produce hydrogen through electrolysis. This allows for the storage and utilization of renewable energy in the form of hydrogen, contributing to the development of a sustainable energy system.

Furthermore, oxygen-ion-conducting SOECs have the potential for co-electrolysis, enabling the simultaneous production of hydrogen and other valuable chemicals or fuels. By adjusting the composition of the feedstock and optimizing the operating conditions, it is possible to produce syngas, methane, or other hydrocarbons directly from carbon dioxide and water. This opens up opportunities for carbon capture and utilization, as well as the production of renewable fuels. While the commercialization of O-SOECs has not been achieved thus far, it is crucial to continue research and development efforts to address any remaining challenges and expedite the commercialization of O-SOECs. This is essential for a cleaner and more sustainable future.

10.4.2 Proton-Conducting SOECs

Proton-conducting solid oxide electrolysis cells (H-SOECs) are a promising technology for efficient and sustainable hydrogen production. These cells operate at high temperatures and utilize a proton-conducting electrolyte, typically made of perovskite-type oxides, to enable the electrolysis of water. For H-SOECs, steam is fed in at the oxygen electrode side to generate oxygen, while hydrogen is produced at the hydrogen electrode side, as shown in Figure 10.5. Compared with O-SOECs with an oxygen-conducting electrolyte, H-SEOCs utilize proton-conducting materials and different principles. In H-SOECs, H ions are transported from the hydrogen electrode to the air electrode while steam is still in the air electrode. Therefore, the hydrogen generated in H-SOECs does not require an additional drying process.

H-SOECs have garnered significant interest due to their advantages, including lower material costs, intermediate to low operating temperatures, higher hydrogen

[8] S. Sahebdelfar, M.T. Ravanchi, Carbon monoxide clean-up of the reformate gas for PEM fuel cell applications: A conceptual review, *Int J Hydrogen Energy*, 48 (2023) 24709–24729.
[9] S. Mucci, A. Mitsos, D. Bongartz, Power-to-X processes based on PEM water electrolyzers: A review of process integration and flexible operation, *Comput Chem Eng*, 175 (2023).
[10] S.S. Kumar, S.U.B. Ramakrishna, B.R. Devi, V. Himabindu, Phosphorus-doped carbon nanoparticles supported palladium electrocatalyst for the hydrogen evolution reaction (HER) in PEM water electrolysis, *Ionics*, 24 (2018) 3113–3121.
[11] M. Gollasch, J. Schmeling, C. Harms, M. Wark, comparative analysis of synthesis routes for antimony-doped tin oxide-supported iridium and iridium oxide catalysts for OER in PEM water electrolysis, *Adv Mater Interfaces*, 10 (2023).
[12] H.Y. Li, Y.X. Xu, N. Lv, Q.Y.T. Zhang, X.L. Zhang, Z.J. Wei, Y.D. Wang, H.L. Tang, H.F. Pan, Ti-doped SnO_2 supports IrO_2 electrocatalysts for the oxygen evolution reaction (OER) in PEM water electrolysis, *ACS Sustain Chem Eng*, 11 (2023) 1121–1132.
[13] P.J. Rheinlander, J. Durst, Transformation of the OER-active IrOx species under transient operation conditions in PEM water electrolysis, *J Electrochem Soc*, 168 (2021).
[14] Y. Wang, Y.H. Pang, H. Xu, A. Martinez, K.S. Chen, PEM Fuel cell and electrolysis cell technologies and hydrogen infrastructure development - A review, Energ *Environ Sci*, 15 (2022) 2288–2328.
[15] M. Carmo, D.L. Fritz, J. Merge, D. Stolten, A comprehensive review on PEM water electrolysis, *Int J Hydrogen Energy*, 38 (2013) 4901–4934.
[16] W. Donitz, E. Erdle, High-temperature electrolysis of water-vapor - Status of development and perspectives for application, *Int J Hydrogen Energy*, 10 (1985) 291–295.

11 Application of *Clostridium Butyricum* to Enhance the in-situ Hydrogen Production and Dechlorination Efficiencies of Chlorinated-Solvent Contaminated Groundwater

Wei-Ting Chen, Jin-Min Li, Jiun-Hau Ou, and Chih-Ming Kao
Institute of Environmental Engineering, National Sun Yat-sen University, Kaohsiung City, Taiwan

Chlorinated hydrocarbons (CHCs) are frequently utilized in industrial manufacturing. Due to their characteristics of being denser than water and having low solubility, they tend to sink and form a plume that spreads with groundwater flow when they infiltrate soil and groundwater. Anaerobic biological degradation is currently considered the most environmentally-friendly remediation technology. The remediation process involves acclimating native organohalide-respiring bacteria (OHRB) or introducing commercial bacterial cultures and ensuring the availability of sufficient electron donors, such as hydrogen (H_2). *Clostridium* carries the *hyd*A gene, which can produce H_2 during the anaerobic fermentation of organic acids, sugars, and alcohols. It also has endospores that can withstand extreme environments and has high hydrogen production efficiency. The organic acids (e.g., acetic acid, butyric acid) in the fermentation products of *Clostridium butyricum* can become an important carbon source for OHRB, which is conducive to the growth of OHRB. As hydrogen accumulates during the fermentation process, under high hydrogen

DOI: 10.1201/9781003463009-11

partial pressure, it can significantly promote the production of organic acids, further improve the hydrogen production efficiency, and cope with the competition for hydrogen from sulfate-reducing bacteria and methanogens during the reductive dechlorination process of OHRB. This leads to the need to provide a more significant amount of hydrogen at the remediation site to reach the threshold value of around 2.2 nM for OHRB reductive dichlorination. Immobilizing *Dehalococcoides mccartyi* BAV1 and *Clostridium butyricum* in a carrier is an innovative approach to enhance the reductive dechlorination of CHCs. The biogenic hydrogen produced by *Clostridium butyricum* can be used for enhanced in-situ reductive dechlorination, which aligns with the goals of green and sustainable remediation more than in-situ hydrogen injection or increasing H_2 levels in water through the reaction of nanoscale zero-valent iron (nZVI) with water. Introducing *Clostridium butyricum* into the remediation site can serve as a potential option for in-situ anaerobic bioremediation of CHCs.

11.1 GROUNDWATER CONTAMINATION BY CHLORINATED SOLVENTS

Due to the rapid development of the chemical industry, many synthetic chemicals have entered the environment through various pathways, altering ecosystems on Earth. Many chemicals commonly used in industry are chalcogenides, such as chlorinated, brominated, or fluorinated hydrocarbons. Among them are chlorinated hydrocarbons (CHCs), in which most hydrogen atoms are replaced by chlorine and have enhanced thermal and chemical stability due to the negatively charged chlorine atoms. These characteristics make them suitable for solvents, herbicides, insecticides, heat transfer fluids, and flame retardants in industrial and agricultural applications (1).

However, improper industrial operations or accidental leaks can result in copious amounts of CHCs being released into the environment. Many CHCs are persistent organic pollutants (2, 3), such as dichloroethylene (DCE), trichloroethylene (TCE), and tetrachloroethylene (PCE), which are often detected in groundwater. When they enter soil and groundwater, their environmental distribution is affected by interactions between media, such as water-air (Harry's Law) and water-soil (adsorption). Due to their higher density than water and low solubility, CHCs are classified as dense non-aqueous phase liquids (DNAPLs). After infiltrating the ground, they sink into the aquifer, due to gravity and capillary action. CHCs are insoluble in water and have high interfacial tension with water, so they form a plume in groundwater and spread irregularly with the flow of groundwater (4–6). This has a significant impact on environmental pollution and human health. To completely dechlorinate CHCs in groundwater, it is crucial to establish a conceptual site model (CSM) for pollutants and develop efficient remediation technologies (7–9).

11.2 IN-SITU BIOREMEDIATION

In-situ biodegradation involves using methods such as bioaugmentation or biostimulation to convert CHCs in groundwater into non-toxic byproducts through microbial reductive dechlorination (10). Biostimulation involves the injection of nutrients (carbon, nitrogen, phosphorus sources, etc.) into the groundwater layer. During the process of utilizing these nutrients, microorganisms also produce substances such as acetate and hydrogen, which serve as the prominent electron donors for reductive dechlorination, stimulating the organohalide-respiring bacteria (OHRB) to perform halogen respiration and achieve reductive dechlorination (11). Studies have shown that adding lactate, ethanol, propionate, butyrate, etc., as substrate donors during remediation can accelerate in-situ reductive dechlorination. Among these, lactate is most commonly used as a carbon source for dechlorinating bacteria because its fermentation byproducts are acetate and propionate, which can stimulate the growth of dechlorinating bacteria and promote reductive dechlorination through co-metabolism (12). The remediation speed of in-situ biostimulation is relatively slow. One of the biggest challenges of this method is how long the nutrient salts that serve as electron donors can remain in the groundwater layer. Therefore, using high molecular weight or slow-release carbon sources (such as vegetable oil or whey) or enhancing in-situ biodegradation through nanoscale zero-valent iron (nZVI) are standard remediation methods (13). Kocur et al. (14) demonstrated that by using carboxymethylcellulose (CMC) to stabilize slow-release carbon sources and nZVI, in-situ reductive dechlorination could be accelerated through a combination of biological and non-biological methods.

Bioaugmentation involves the addition of pre-cultured and well-grown dechlorinating bacteria to a contaminated site to enhance the degradation of target pollutants. The selected microorganisms usually have complete dechlorination genes and better environmental tolerance (15). Many commercial products, such as KB-1™ (SiREM; http://www.siremlab.com/) and SDC-9 (Shaw Environmental Inc., Lawrenceville, New Jersey, USA), and these commercial bacterial agents (such as those shown in Table 11.1)

TABLE 11.1
Commonly Used Commercial Microbial Agents in Recent Years

Augmented Microorganisms	Target	Sediment/Soil	Electron Donor/Carbon Source	Degradation Activities	References
SDC-9 (containing *Dehalococcoides*)	PCE	Edwards air force base	Lactate	Ethene	(18)
KB-1	TCE	TCE contaminated, water-bearing conglomerate site near Tracy, California	Lactate	Ethene	(17)
KB-1	TCE	Southern Ontario (ISSO), Canada	Ethanol	Ethene	(12)
Anaerobic activated sludge collected from an industrial wastewater-treatment plant	TCE	Southern Taiwan	Brown sugar	Cis-DCE/ Vinyl chloride/ Ethene	(25)

almost all contain multiple strains of *Dehalococcoides* (*Dhc*) (12, 16–18). Dutta, Thomsen (19) treated 100 mg/L of TCE with KB1 combined with a slow-release carbon source and achieved a conversion rate of 80% for TCE and 79–83% for cis-1,2-DCE and trans-1,2-DCE within 200 days of the experiment, with 63% of the conversion to ethane (Eth) in the remediation system. Kucharzyk et al. (20) confirmed that the commercial bacterial agent SDC-9 containing *Dhc* was able to identify 12 reductive dehalogenase (RDase) genes, including *pceA*, *vcrA*, and *tceA*. Michalsen et al. (21) pointed out that SDC-9 played a vital role in the reductive dechlorination rate of cis-1,2-DCE and VC. However, when bioaugmentation is applied to in-situ bioremediation to remove CHCs, it may compete with native bacteria or result in insufficient nutrients, leading to poor remediation results. Therefore, combining biostimulation and bioaugmentation is a more appropriate remediation method (9).

Biostimulation and bioaugmentation are often combined in permeable reactive barriers (PRBs) for in-situ treatment of soil and groundwater contaminated with CHCs. A PRB is formed by dispersing a medium that can treat or promote the growth of CHC-degrading bacteria in the groundwater layer. When groundwater flows through the reactive zone, CHCs are degraded by the materials or bacteria in the medium. Currently, the mainstream approach is to inject slow-release carbon sources such as long-lasting emulsified colloidal substrates to acclimate native dechlorinating bacteria, and commercial bacterial culture solution are injected as needed to enhance biodegradation (22–24).

11.3 GREEN AND SUSTAINABLE REMEDIATION

Microbial reductive dechlorination is recognized as a green and sustainable remediation method. The remediation process requires an understanding of dehalogenation bacteria's growth (OHRB) and environmental control. In addition to the presence of OHRB on site, the supplementation of nutrients is also one of the means to enhance bioremediation (19). Chen et al. (26) developed an emulsified castor oil substrate that can adsorb TCE and provide nutrients required for biodegradation, removing 97% of TCE within 95 days of the experiment. Sheu et al. (27) developed a long-lasting emulsified substrate containing vegetable oil, surfactants, and nZVI that can maintain a total organic carbon concentration in water up to 488 mg/L, removing 99% of TCE (initial TCE concentration 7.4 mg/L) within 133 days. nZVI played a crucial role in preventing groundwater acidification and inhibiting the biological inhibition and odor caused by H_2S production during remediation.

Wang et al. (28) mentioned that in-situ bioremediation is an effective solution for dealing with persistent halogenated organic pollutants. Many remediation agents developed by researchers provide nutrients required for the growth of OHRB but also produce hydrogen through biodegradation, which is an important mechanism. Chlorinated organic compounds such as PCE and TCE in groundwater have high oxidation states and can be converted into cis-1,2-dichloroethylene (cis-1,2-DCE), vinyl chloride (VC), and Eth in anaerobic environments. Since cis-1,2-DCE and VC are low oxidation state compounds, degradation under anaerobic conditions requires more energy (9). OHRB use H_2 or other organic matter as electron donors to achieve reductive dechlorination through reductive dehalogenase (RDase). Common RDases include *pce*A, *tec*A, *bvc*A, and *vcr*A. *pce*A leads to the reductive dechlorination of PCE to TCE, while *tce*A can degrade TCE or DCE to VC. *bvc*A and *vcr*A can degrade species below dichloronation to Eth and are indicator genes for whether complete reductive dechlorination can be achieved on-site (10). The related reaction pathways are shown in Figures 11.1 and 11.2.

The main factor affecting the efficiency of reductive dechlorination is the competition between OHRB and other microorganisms for electron donors in the environment under anaerobic conditions. H_2 is an essential electron donor for the dechlorination of *cis*-1,2-DCE and VC to produce Eth. Methanogens, homoacetogens, and sulfidogens in the environment compete with OHRB for hydrogen. In this competitive relationship, the demand for hydrogen in water is usually higher than the theoretical demand for simple dechlorination. Therefore, to achieve complete dechlorination, in addition to paying attention to OHRB in water, it is also necessary to ensure sufficient hydrogen as electron donors (28–30). Previous studies have shown that the hydrogen threshold required by OHRB is approximately 2.2 nM (31).

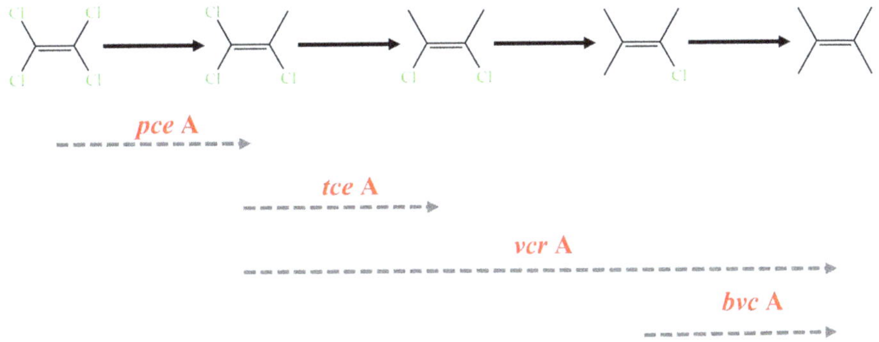

FIGURE 11.1 Microbial reductive dechlorination pathway and the involvement of RDase.

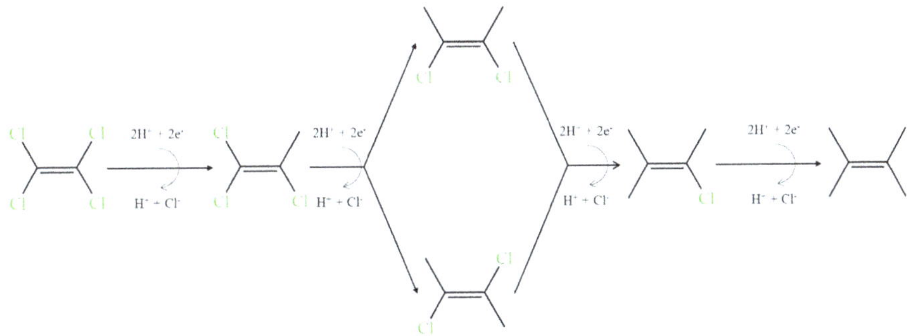

FIGURE 11.2 The reductive dechlorination pathway of chlorinated organic compounds.

11.4 HYDROCARBON-PRODUCING BACTERIA AND *CLOSTRIDIUM BUTYRICUM*

H_2 is one of the products produced by many microorganisms during fermentation, such as *Clostridium* spp., *Bacillus* spp., and *Enterobacter* spp., which have been widely used in fermentative hydrogen production. Among them, Clostridium spp. is known for its high hydrogen production efficiency and ability to utilize a wide range of carbon sources (32, 33). Clostridium is an anaerobic, endospore-forming, Gram-positive bacterium widely found in soil and groundwater and highly tolerant to extreme environments. It can quickly become the dominant genus in microbial communities through simple cultivation. The optimal pH conditions for Clostridium are pH 5.0–9.0, and the optimal growth temperature is <37°C (33).

Karube et al. (34) were the first to use *Clostridium butyricum* for fermentative hydrogen production. Due to its high hydrogen production and ease of cultivation, it has gradually become a common strain for fermentative hydrogen production. Sim et al. (35) summarized the carbon sources that can be utilized by *Clostridium butyricum*, including carbohydrates such as glucose, fructose, lactose, xylose, starch, hexose, and glycerol, with a hydrogen production efficiency of 0.23–3.57 moles H_2/mole hexose. RamKumar et al. (36) used *Clostridium butyricum* TM-9A in a 13.5 L bioreactor, with molasses as the carbon source for fermentation, achieving a high hydrogen production rate of 46–73 mmol/L. Kim et al. (37) found that using conductive metals and lactate can improve the efficiency of hydrogen production, and the addition of lactate is a critical factor in enhancing hydrogen production. During fermentation, the activation of hydrogenase (*hyd*A) can be observed. Many studies have shown that *Clostridium butyricum* can achieve high hydrogen production efficiency by adding appropriate carbon sources, demonstrating its potential for flexible application (35, 38).

11.5 ENHANCED DECHLORINATION BY *CLOSTRIDIUM BUTYRICUM*

OHRB requires H_2 as an electron donor for reductive dechlorination, having a continuous and stable source of hydrogen is critical to anaerobic reductive dechlorination. As large amounts of carbon sources are often injected during on-site remediation, microorganisms such as *Clostridium, Syntrophomonas*, and *Enterobacter*, which carry the *hyd*A gene, can produce H_2 during anaerobic fermentation of organic acids, sugars, and alcohols. Using biogenic hydrogen produced by microorganisms is more in line with green and sustainable remediation goals than in-situ hydrogen injection or increasing H_2 levels in water through the reaction of nZVI with water (39, 40). Among them, *Clostridium* is currently the most studied hydrogen-producing bacterium. Therefore, using hydrogen-producing bacteria to provide a stable and sufficient hydrogen supply to enhance in-situ reductive dechlorination is feasible.

Lu et al. (41) enhanced the reductive dechlorination of TCE and *cis*-1,2-DCE by immobilizing *Dehalococcoides mccartyi* BAV1 and *Clostridium butyricum* in a silica. Their study observed degradation byproducts such as 1,1-DCE, cis-1,2-DCE, VC, and Eth. In the TCE experimental group with the addition of silicone particles containing immobilized hydrogen-producing bacteria, 69.24% of TCE (initial TCE concentration 2.5 mg/L) was degraded within 30 days, and the number of *Dehalococcoides mccartyi* BAV1 increased significantly by at least 10 times. In the cis-1,2-DCE experiment, it was observed that immobilizing *Dehalococcoides mccartyi* BAV1 and *Clostridium butyricum* together in a silicone carrier could indeed completely dechlorinate cis-1,2-DCE. Since Clostridium butyricum can provide H2 for a long time to maintain the activity of *Dehalococcoides mccartyi* BAV1, Eth generation can be observed throughout the 180-day experiment, indicating that *Clostridium* has a high potential to promote the reductive dechlorination of TCE and other compounds and can solve the problem of VC accumulation in reductive dechlorination reactions.

11.6 CASE STUDIES

In Lo et al. (42) study. An innovative immobilized *Clostridium butyricum* (ICB) column scheme utilizing hydrogen-producing bacteria was developed and applied to remediate groundwater polluted with chlorinated ethenes, primarily cis-1,2-dichloroethene (cis-DCE), through anaerobic reductive dechlorination processes (Figure 11.3). The objectives were to evaluate the effectiveness of the ICB scheme in cleaning up cis-DCE polluted groundwater and to characterize changes in microbial communities following ICB application. Three remediation wells and two monitoring wells were installed within the cis-DCE plume reaction well, a 1.2-m PVC column with a 2.5 cm radius filled with ICB beads and 20 L of the slow polycolloid-releasing substrate (SPRS) to enhance hydrogen production and provide primary carbon supply, respectively. Groundwater samples from the remediation and monitoring wells were periodically analyzed for cis-DCE and its degradation by products, microbial diversity, reductive dehalogenase, and geochemical indicators. Results showed a significant decrease in cis-DCE within the influence zone of ICB and SPRS. In a remediation well with ICB injection, approximately 98.4% of cis-DCE removal (initial concentration ration = 1.46 mg/L) was observed with

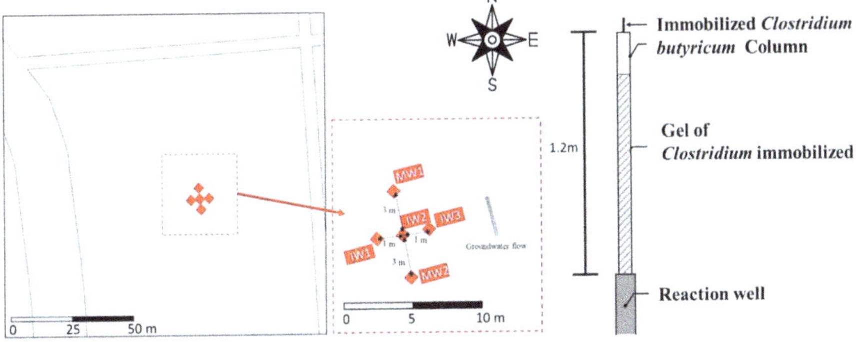

FIGURE 11.3 Enhanced reductive dechlorination pilot test using silica gel particles immobilized with *Clostridium butyricum*.

the production of Eth (the end-product of cis-DCE dechlorination) after 56 days of system operation. Up to 0.72 mg/L of hydrogen was observed in remediation wells 14 days after the introduction of ICB and SPRS, corresponding to an increased population of *Dhc* (increased from 3.76×10^3 to 5.08×10^5 gene copies/L). Metagenomics analysis revealed that introducing SPRS and ICB significantly impacted bacterial com munities, with increased *Bacteroides*, *Citrobacter*, and *Desulfovibrio* observed, contributing significantly to *cis*-1,2-DCE's reductive dechlorination. ICB application effectively increased *Dhc* and RDase gene populations, corresponding to improved dechlorination of *cis*-1,2-DCE and VC. The introduction of ICB and SPRS could serve as a potential in-situ remedial option to enhance the anaerobic dechlorination efficiencies of chlorinated ethenes.

REFERENCES

1. Wang S, Chen S, Wang Y, Low A, Lu Q, Qiu R. Integration of organohalide-respiring bacteria and nanoscale zero-valent iron (Bio-nZVI-RD): A perfect marriage for the remediation of organohalide pollutants? *Biotechnology Advances*. 2016;34(8):1384–95.
2. Lin X-Q, Li Z-L, Liang B, Zhai H-L, Cai W-W, Nan J, et al. Accelerated microbial reductive dechlorination of 2, 4, 6-trichlorophenol by weak electrical stimulation. *Water Research*. 2019;162:236–45.
3. Ji C, Meng L, Wang H. Enhanced reductive dechlorination of 1, 1, 1-trichloroethane using zero-valent iron-biochar-carrageenan microspheres: preparation and microcosm study. *Environmental Science and Pollution Research*. 2019;26:30584–95.
4. Eklund B, Rago R, Plantz G, Haddad E, Miesfeldt M, Volpi R. Fate & transport of vinyl chloride in soil vapor. *Remediation Journal*. 2022;32(4):273–9.
5. Kaown D, Lee S-S, Lee KK, editors. (2022, December). Unravelling sources and fate of chlorinated solvents in groundwater at an industrial area using compound specific isotope and microbial data. In AGU Fall Meeting Abstracts (Vol. 2022, pp. H25H-1207).
6. Lee S-S, Jun S-C, Ha S-W, Lee S, Lee KK, editors. Evaluating the fate and transport of chlorinated ethenes in an industrial complex using the combined long-term monitoring results. AGU Fall Meeting Abstracts; 2021.
7. Usman M, Shi Z, Dutta N, Ashraf MA, Ishfaq B, El-Din MG. Current challenges of hydrothermal treated wastewater (HTWW) for environmental applications and their perspectives: A review. *Environmental Research*. 2022;212:113532.
8. Dutta N, Usman M, Ashraf MA, Luo G, Zhang S. A critical review of recent advances in the bio-remediation of chlorinated substances by microbial dechlorinators. *Chemical Engineering Journal Advances*. 2022;12:100359.
9. Xiao Z, Jiang W, Chen D, Xu Y. Bioremediation of typical chlorinated hydrocarbons by microbial reductive dechlorination and its key players: A review. *Ecotoxicology and Environmental Safety*. 2020;202:110925.
10. Saiyari DM, Chuang H-P, Senoro DB, Lin T-F, Whang L-M, Chiu Y-T, et al. A review in the current developments of genus Dehalococcoides, its consortia and kinetics for bioremediation options of contaminated groundwater. *Sustainable Environment Research*. 2018;28(4):149–57.
11. Scheutz C, Durant ND, Dennis P, Hansen MH, Jørgensen T, Jakobsen R, et al. Concurrent ethene generation and growth of Dehalococcoides containing vinyl chloride reductive dehalogenase genes during an enhanced reductive dechlorination field demonstration. *Environmental Science & Technology*. 2008;42(24):9302–9.

12. Perez-de-Mora A, Zila A, McMaster ML, Edwards EA. Bioremediation of chlorinated ethenes in fractured bedrock and associated changes in dechlorinating and nondechlorinating microbial populations. *Environmental Science & Technology*. 2014;48(10):5770–9.
13. Matturro B, Pierro L, Frascadore E, Petrangeli Papini M, Rossetti S. Microbial community changes in a chlorinated solvents polluted aquifer over the field scale treatment with poly-3-hydroxybutyrate as amendment. *Frontiers in Microbiology*. 2018;9:1664.
14. Kocur CMD, Lomheim L, Molenda O, Weber KP, Austrins LM, Sleep BE, et al. Long-term field study of microbial community and dechlorinating activity following carboxymethyl cellulose-stabilized nanoscale zero-valent iron injection. *Environmental Science & Technology*. 2016;50(14):7658–70.
15. Dolinová I, Štrojsová M, Černík M, Němeček J, Macháčková J, Ševců A. Microbial degradation of chloroethenes: A review. *Environmental Science and Pollution Research*. 2017;24:13262–83.
16. Adetutu EM, Gundry TD, Patil SS, Golneshin A, Adigun J, Bhaskarla V, et al. Exploiting the intrinsic microbial degradative potential for field-based in-situ dechlorination of trichloroethene contaminated groundwater. *Journal of Hazardous Materials*. 2015;300:48–57.
17. Verce MF, Madrid VM, Gregory SD, Demir Z, Singleton MJ, Salazar EP, et al. A long-term field study of in-situ bioremediation in a fractured conglomerate trichloroethene source zone. *Bioremediation Journal*. 2015;19(1):18–31.
18. Schaefer CE, Lavorgna GM, White EB, Annable MD. Bioaugmentation in a well-characterized fractured rock DNAPL source area. *Groundwater Monitoring & Remediation*. 2017;37(2):35–42.
19. Dutta N, Thomsen K, Ahring BK. Degrading chlorinated aliphatics by reductive dechlorination of groundwater samples from the Santa Susana Field Laboratory. *Chemosphere*. 2022;298:134115.
20. Kucharzyk KH, Meisel JE, Kara-Murdoch F, Murdoch RW, Higgins SA, Vainberg S, et al. Metagenome-guided proteomic quantification of reductive dehalogenases in the dehalococcoides mccartyi-containing consortium SDC-9. *Journal of Proteome Research*. 2020;19(4):1812–23.
21. Michalsen MM, Kara Murdoch F, Löffler FE, Wilson J, Hatzinger PB, Istok JD, et al. Quantitative proteomics and quantitative PCR as predictors of cis-1,2-dichlorethene and vinyl chloride reductive dechlorination rates in bioaugmented aquifer microcosms. *ACS ES&T Engineering*. 2022;2(1):43–53.
22. Mondal PK, Lima G, Zhang D, Lomheim L, Tossell RW, Patel P, et al. Evaluation of peat and sawdust as permeable reactive barrier materials for stimulating in-situ biodegradation of trichloroethene. *Journal of Hazardous Materials*. 2016;313:37–48.
23. Bekele DN, Du J, de Freitas LG, Mallavarapu M, Chadalavada S, Naidu R. Actively facilitated permeable reactive barrier for remediation of TCE from a low permeability aquifer: Field application. *Journal of Hydrology*. 2019;572:592–602.
24. Thakur AK, Vithanage M, Das DB, Kumar M. A review on design, material selection, mechanism, and modelling of permeable reactive barrier for community-scale groundwater treatment. *Environmental Technology & Innovation*. 2020;19:100917.
25. Chiu H, Liu J, Chien H, Surampalli R, Kao C. Evaluation of enhanced reductive dechlorination of trichloroethylene using gene analysis: pilot-scale study. *Journal of Environmental Engineering*. 2013;139(3):428–37.
26. Chen W-T, Chen K-F, Surmpalli RY, Zhang TC, Ou J-H, Kao C-M. Bioremediation of trichloroethylene-polluted groundwater using emulsified castor oil for slow carbon release and acidification control. *Water Environment Research*. 2022;94(1):e1673.

27. Sheu YT, Chen SC, Chien CC, Chen CC, Kao CM. Application of a long-lasting colloidal substrate with pH and hydrogen sulfide control capabilities to remediate TCE-contaminated groundwater. *Journal of Hazardous Materials*. 2015;284:222–32.
28. Wang S, Qiu L, Liu X, Xu G, Siegert M, Lu Q, et al. Electron transport chains in organohalide-respiring bacteria and bioremediation implications. *Biotechnology Advances*. 2018;36(4):1194–206.
29. Lin WH, Chien CC, Lu CW, Hou D, Sheu YT, Chen SC, et al. Growth inhibition of methanogens for the enhancement of TCE dechlorination. *Science of the Total Environment*. 2021;787:147648.
30. Lin W-H, Chen C-C, Sheu Y-T, Tsang DCW, Lo K-H, Kao C-M. Growth inhibition of sulfate-reducing bacteria for trichloroethylene dechlorination enhancement. *Environmental Research*. 2020;187:109629.
31. Dutta N, Usman M, Ashraf MA, Luo G, Zhang S. Efficacy of emerging technologies in addressing reductive dechlorination for environmental bioremediation: A review. Journal of Hazardous *Materials Letters*. 2022:100065.
32. Castelló E, Ferraz-Junior ADN, Andreani C, del Pilar Anzola-Rojas M, Borzacconi L, Buitrón G, et al. Stability problems in the hydrogen production by dark fermentation: possible causes and solutions. *Renewable and Sustainable Energy Reviews*. 2020; 119:109602.
33. Wang J, Yin Y. Progress in microbiology for fermentative hydrogen production from organic wastes. *Critical Reviews in Environmental Science and Technology*. 2019;49(10):825–65.
34. Karube I, Matsunaga T, Tsuru S, Suzuki S. Continous hydrogen production by immobilized whole cells of Clostridium butyricum. *Biochimica et Biophysica Acta (BBA)-General Subjects*. 1976;444(2):338–43.
35. Sim Y-B, Yang J, Kim SM, Joo H-H, Jung J-H, Kim D-H, et al. Effect of bioaugmentation using Clostridium butyricum on the start-up and the performance of continuous biohydrogen production. *Bioresource Technology*. 2022;366:128181.
36. RamKumar N, Anupama PD, Nayak T, Subudhi S. Scale up of biohydrogen production by a pure strain; Clostridium butyricum TM-9A at regulated pH under decreased partial pressure. *Renewable Energy*. 2021;170:1178–85.
37. Kim D-H, Yoon J-J, Kim S-H, Park J-H. Acceleration of lactate-utilizing pathway for enhancing biohydrogen production by magnetite supplementation in Clostridium butyricum. *Bioresource Technology*. 2022;359:127448.
38. Arisht SN, Roslan R, Gie GA, Mahmod SS, Sajab MS, Lay C-H, et al. Effect of nano zero-valent iron (nZVI) on biohydrogen production in anaerobic fermentation of oil palm frond juice using Clostridium butyricum JKT37. *Biomass and Bioenergy*. 2021;154:106270.
39. Wang J, Ng CK, Cao B, Qing W, Liu F, Tang CY. Polydopamine enabled palladium loaded nanofibrous membrane and its catalytic performance for trichloroethene dechlorination. *Applied Catalysis A: General*. 2018;559:122–6.
40. Mohammed O, Mumford KG, Sleep BE. Effects of hydrogen gas production, trapping and bubble-facilitated transport during nanoscale zero-valent iron (nZVI) injection in porous media. *Journal of Contaminant Hydrology*. 2020;234:103677.
41. Lu C-W, Kao C-M, Le NN, Lin C-C, Chen S-C. Long-term dechlorination of cis-DCE to ethene with co-immobilized Dehalococcoides mccartyi BAV1 and Clostridium butyricum in silica gel system. *Journal of Hazardous Materials*. 2022;430:128355.
42. Lo K-H, Lu C-W, Chien C-C, Sheu Y-T, Lin W-H, Chen S-C, et al. Cleanup chlorinated ethene-polluted groundwater using an innovative immobilized Clostridium butyricum column scheme: A pilot-scale study. *Journal of Environmental Management*. 2022;311: 114836.

12 Recycling Batteries Materials

Wei-Sheng Chen and Cheng-Han Lee
National Cheng Kung University, Tainan City, Taiwan

12.1 BACKGROUND

12.1.1 Introduction of Battery

A battery is an electrochemical device that stores chemical energy and converts it into electrical energy [1]. Through chemical reactions within an electrochemical cell, electrons flow from one electrode to another, generating an electric current that can be harnessed for various purposes. A battery can comprise one or multiple cells, essentially serving as compact chemical reactors where energetic electrons are produced and poised to traverse the external circuit [2]. The origins of the first authentic battery can be traced back to 1800 when Alessandro Volta, an Italian physicist, invented it. The invention of the battery as we know it is credited to Volta, who put together the first battery to prove a point to another Italian scientist, Luigi Galvani. Galvani had shown that the legs of frogs hanging on iron or brass hooks would twitch when touched with a probe of some other type of metal. He believed this was caused by electricity from within the frogs' tissues and called it 'animal electricity'. Batteries rely on diverse chemical compositions that form the fundamental unit known as a cell. Various battery types exist, encompassing wet cells and dry cells. Wet cells consist of liquid electrolytes, whereas dry cells contain absorbent material to retain the electrolyte (Table 12.1). Multiple batteries can be connected in series or parallel, effectively augmenting the total voltage or current output, respectively.

The common classification of battery is primary battery and secondary battery. A primary battery, also known as a non-rechargeable battery, is a type of battery that cannot be recharged once it has been fully discharged. Primary batteries are designed for single-use applications and are typically discarded after depleting their energy. These batteries generate electrical energy through a chemical reaction within the battery cells. Common examples of primary batteries include alkaline batteries, zinc-carbon batteries, and mercury batteries. On the other hand, a secondary battery, also known as a rechargeable battery, can be recharged multiple times by reversing the chemical reactions that occur during discharge. Secondary batteries are designed for repeated use and can be recharged using an external power source, such as an electrical outlet or a dedicated battery charger. This allows them to regain their energy and be used again. Common examples of secondary batteries include lithium-ion (Li-ion) batteries, nickel-metal hydride (NiMH) batteries, and lead-acid batteries. The applications of the two batteries are shown below.

DOI: 10.1201/9781003463009-12

TABLE 12.1
Difference between Wet Cells and Dry Cells

	Size	Electrolyte	Leaking	Operation	Manufacture	Overcharge
Wet Cells	Large	Liquid	Easy	Difficult	Easy	Can
Dry Cells	Small	Moist Solid	Difficult	Easy	Difficult	Can not

Source: https://pediaa.com/difference-between-dry-cell-and-wet-cell/

Primary Battery Applications:

1. Portable Electronics: Primary batteries are widely used in portable electronic devices such as remote controls, calculators, watches, cameras, and small appliances.
2. Emergency Equipment: They are commonly used in emergency preparedness kits, flashlights, radios, and smoke detectors, providing reliable power during power outages or emergencies.
3. Medical Devices: Primary batteries are used in medical devices such as hearing aids, glucose meters, thermometers, and various types of monitoring equipment.
4. Toys and Games: Many battery-operated toys, handheld gaming devices, and remote-controlled vehicles rely on primary batteries for power.
5. Transportation: Primary batteries find applications in transportation devices like keyless entry systems, vehicle remote controls, and tire pressure monitoring systems.

Secondary Battery Applications:

1. Electric Vehicles (EVs): Secondary batteries, particularly lithium-ion batteries, are commonly used in electric cars, hybrid vehicles, and electric bicycles due to their high energy density and rechargeability.
2. Portable Electronics: Secondary batteries power various portable electronic devices like smartphones, laptops, tablets, e-readers, and portable gaming consoles.
3. Renewable Energy Systems: Secondary batteries are used to store energy generated from renewable sources such as solar panels and wind turbines, allowing for a continuous power supply even when the source is unavailable or intermittent.
4. Power Tools: Cordless power tools, such as drills, saws, and sanders, often utilize secondary batteries for mobility and convenience on construction sites or other remote locations.
5. Backup Power Supplies: Secondary batteries can provide backup power for critical systems and equipment, such as computer servers, telecommunications equipment, and emergency lighting systems.

12.1.2 The Current Situation of Waste Batteries

As abovementioned, batteries are now broadly employed in many areas. Thus, the number of waste batteries is increased as well. There are several reasons why there are so many waste batteries:

1. Increased consumption: With the advancement of technology, the demand for battery-powered devices has risen significantly. People use batteries in various electronic devices such as smartphones, laptops, tablets, digital cameras, remote controls, toys, and many more. This increased consumption leads to a higher number of batteries being discarded.
2. Single-use batteries: Many batteries used in everyday items are disposable or single-use batteries, such as alkaline batteries. These batteries are not designed to be recharged and are discarded after their energy is depleted. The use of single-use batteries contributes to the accumulation of battery waste.
3. Limited recycling infrastructure: While batteries can be recycled, the infrastructure for battery recycling is limited in many places. Not all communities have convenient battery recycling programs or facilities. As a result, many people throw batteries in the trash, leading to increased battery waste.
4. Lack of awareness: Some people are not aware of the environmental impact of improper battery disposal or the availability of recycling options. They may not realize that batteries contain hazardous substances such as heavy metals (e.g., mercury, lead, cadmium) that can contaminate soil and water if not disposed of properly.
5. Global trends and e-waste: The rise in electronic waste (e-waste) also contributes to the increasing number of waste batteries. When electronic devices are discarded, their batteries are often not removed or disposed of separately. This leads to batteries mixed with other e-waste and ending up in landfills.

According to statistics, Americans throw away more than three billion batteries yearly. This amounts to about 180,000 tons of waste batteries annually [3]. However, it is important to note that this statistic only applies to the United States. On the other hand, the European Commission declares that approximately 800,000 tons of automotive batteries, 190,000 tons of industrial batteries, and 160,000 tons of consumer batteries enter the European Union every year [4, 5]. To address the issue of waste batteries, it is important to promote battery recycling programs, improve public awareness about proper disposal methods, and encourage the development of more sustainable battery technologies. Additionally, efforts are being made to improve battery recycling infrastructure and explore alternative solutions like battery leasing and sharing models to reduce overall battery consumption. The paragraphs below will introduce the recycling techniques of two primary batteries (zinc-carbon and mercury batteries) and two secondary batteries (lithium-ion and lead-acid batteries).

12.2 RECYCLING TECHNIQUES OF ZINC-CARBON BATTERIES

A zinc-carbon battery is a primary dry cell that generates direct electric current through the electrochemical reaction between zinc (Zn) and manganese dioxide (MnO_2) in the presence of an electrolyte composed of ammonium chloride (NH_4Cl). It emerged as a dry battery variant of the Leclanché cell and became the first commercially available dry battery [6, 7]. The zinc/carbon cell uses a zinc anode and a manganese dioxide cathode, and the carbon is added to the cathode to increase conductivity and retain moisture. The overall reaction in the cell is $Zn + 2MnO_2 \rightarrow ZnO + Mn_2O_3$. Its introduction enabled the development of portable devices like flashlights, as it offered a higher energy density at a more affordable price compared to existing cells. Even today, zinc-carbon batteries find utility in low-power or sporadically used gadgets such as remote controls, flashlights, clocks, and transistor radios.

According to the United States Federal definition of solid waste, waste carbon zinc and standard alkaline batteries are considered solid waste. Zinc-carbon and alkaline dry cell batteries constitute approximately 80% of the waste dry-cell battery stream. Take Taiwan as an instance, about 966 tons of waste dry-cell batteries are produced by industries, but all are transported overseas for treatment. Therefore, it needs to conduct recycling techniques to save costs and extract resources from zinc-carbon batteries.

Nowadays, the recycling techniques of waste zinc-carbon batteries are mainly focused on recovering metals. Ferella F et al. proposed a recycling procedure to recover zinc and manganese from spent alkaline and zinc–carbon batteries [8]. In the laboratory, a series of tests are conducted to produce a refined solution containing valuable zinc (with a purity of 99.6%) that can be retrieved through electrolysis. Simultaneously, manganese is recovered in the form of oxide mixtures by subjecting the solid residue obtained from the leaching process to high-temperature roasting. After 3 hours at 80 °C, using a pulp density of 10% w/v and a sulphuric acid concentration of 1.5 M, approximately 99% of zinc and 20% of manganese are successfully extracted. To purify the leach solution, iron is selectively precipitated, while unwanted metallic impurities like copper, nickel, and cadmium are eliminated through cementation with zinc powder. The remaining solid residue from the leaching stage is then subjected to a 30-minute roasting at 900 °C, completely removing graphite and yielding a mixture of Mn_3O_4 and Mn_2O_3 with a manganese grade of 70%.

Shin S M et al. put forward ideas to selectively separate zinc [9]. To separate non-magnetic components (88.4 wt.%) from spent zinc-carbon batteries, a combination of physical techniques, including crushing, sieving, and magnetic separation, was employed. The larger particles obtained were then subjected to eddy current separation, facilitating the recovery of zinc sheets, carbon rods, and plastics. The smaller fraction (below 2.36 mm) containing 15.5% Zn, 17.5% Mn, and 1.4% Fe was utilized for leaching experiments conducted under varying conditions such as sodium hydroxide concentration, temperature, agitation speed, and pulp density. By implementing selective leaching with a solid/liquid ratio of 100 g dm^{-3}, a sodium hydroxide concentration of 4 mol dm^{-3}, temperature of 80 °C, and agitation speed of 200 rpm, a zinc extraction rate of 82% and a manganese extraction rate below 0.1% were achieved. Consequently, the processes outlined in this study led to an overall zinc recovery rate of approximately 88.5%.

Khan M H et al. employed two different ways to obtain valuable metals from waster zinc-carbon batteries [10]. Subjecting the anodes to a heating process at 600 °C for 10 minutes in the presence of a 12% NH4Cl flux made it possible to extract a maximum of 92% of the total zinc content with a purity exceeding 99.0%. In the case of spent electrolyte paste containing manganese and zinc as the primary metallic elements, leaching was conducted using a sulfuric acid solution, along with hydrogen peroxide as a reducing agent. Optimal leaching conditions were determined as follows: sulfuric acid concentration of 2.5 M, hydrogen peroxide concentration of 10%, temperature of 60 °C, stirring speed of 600 rpm, and a solid/liquid ratio of 1:12. Under these optimized conditions, up to 88% of the manganese present in the paste could be dissolved within 27 minutes of leaching. Similarly, the dissolution of zinc under the same conditions reached 97%. Subsequently, precipitation of the leach liquor allowed for the recovery of 69.89% of manganese and 83.29% of zinc in the form of manganese carbonate and zinc oxalate.

In addition to separating and obtaining metals, some scholars turned the metals into composite materials. To reuse spent batteries on a large scale, this study concerns a simple, effective, and sustainable strategy to turn them into MnO/ZnO/C composites [11]. By subjecting rust cathode materials to conventional leaching treatment followed by pyrolysis, the resulting product is a composition of MnO/ZnO/C. When utilized as a cathode in rechargeable zinc-ion batteries, this electrode exhibits a maximum reversible capacity of approximately 362 mAh g^{-1} MnO and a rate capability of 191 mAh g^{-1} MnO, even at a high current rate of 1.20 A g^{-1}. Additionally, the ZnO component gradually dissolves in the electrolyte over discharge cycles, effectively replenishing the Zn^{2+} content and thereby improving cycling stability (98.02% after 500 cycles). The device also showcases a notable energy density of 336.37 Wh kg^{-1}, demonstrates a low self-discharge rate, and efficiently powers an LED panel. This approach presents an economical and straightforward method for converting zinc-carbon battery waste into valuable materials suitable for aqueous rechargeable zinc-ion batteries.

Based on the above narrative, it can be noticed that leaching is a common process in the recycling techniques of waste zinc-carbon batteries. Leaching is a process that involves the use of a solvent to remove a solute from a solid mixture [12]. The process can be used to extract valuable metals from ores, remove contaminants from waste, and separate valuable materials from waste products. Leaching is a mass transfer process that takes place by extracting a substance from a solid material that has come into contact with the liquid. The leaching process generally concerns processes where the solid is inert and contains soluble solute extracted from the inert solid with the help of a chemical reaction. Leaching is a technique which belongs to hydrometallurgy. Apart from leaching, other hydrometallurgy techniques, such as chemical precipitation, solvent extraction, and ion-exchange, are also common in separating and purifying metals from wastes [12].

12.3 RECYCLING TECHNIQUES OF MERCURY BATTERIES

A mercury battery, also referred to as a mercuric oxide battery, mercury cell, button cell, or Ruben-Mallory battery, is a primary electrochemical cell that cannot be recharged [13]. These batteries operate through a chemical reaction between mercuric oxide and zinc electrodes immersed in an alkaline electrolyte. During discharge, the voltage of a mercury battery remains almost constant at 1.35 volts, while offering a significantly higher capacity compared to similarly sized zinc-carbon batteries. Button cell versions of mercury batteries were commonly used in devices like watches, hearing aids, cameras, and calculators while larger variants were employed in various other applications. The cathode of a mercury battery employs either pure mercuric oxide (HgO) or a mixture of HgO with manganese dioxide (MnO_2). Since mercuric oxide is not conductive, a small amount of graphite is added to the cathode mixture, which also helps prevent the formation of large mercury droplets. The half-reaction occurring at the cathode possesses a standard potential of +0.0977 V [14].

Although mercury batteries can be applied in many areas, it contains mercury which may pollute water and soil. Early mercury batteries have been banned due to environmental pollution issues. However, other button cell batteries may still contain small amounts of mercury. Moreover, although manufacturers have developed mercury-free alkaline batteries and silver oxide batteries in recent years, various countries still allow a limited amount of mercury in batteries due to technological and patent-related reasons. Therefore, if waste mercury batteries are stored or treated inappropriately, it will not only contaminate the environment but also pass along the food chain and eventually to humans. Besides, multiple studies have examined the correlation between child health, autism spectrum disorder (ASD), and heavy metals like mercury. The findings indicate that exposure to various sources of contamination, including polluted air, water, soil, and food, may contribute to the development of behavioral disorders in children. Additionally, exposure to toxic metals during pregnancy has been linked to an increased risk of ASD in children, whether it occurs before or after birth [15–17]. Furthermore, certain research has explored the association between mercury levels and diabetes. The results suggest that higher mercury exposure during early adulthood may raise the likelihood of developing diabetes later in life [18, 19]. Based on the above narrative, it can be understood that mercury treatment is critical. Therefore, the recycling techniques of waste mercury batteries should be implemented to avoid mercury leaking.

As mentioned above, mercury batteries should be recycled instead of being dumped in landfills or incinerated, as they can leak mercury toxins into water supplies and food chains. The recycling process for mercury batteries involves liquid and heat extraction methods, which must be done in controlled environments due to the presence of highly toxic heavy metals. Here are the steps involved in battery recycling:

1. Waste Preparation: The battery components are separated into two end products through a mechanical separation process at room temperature. Recyclers separate plastic materials from metal components, and both materials are perfect for making new products.
2. Metallurgical Processing: The components of zinc-based batteries undergo a process known as High-Temperature Metal Reclamation, where metals like zinc, manganese, chromium, and iron are extracted and used to make new products.
3. Mercury Battery Recycling: Mercury batteries undergo recycling through heat extraction methods (pyrometallurgy), which must be conducted in controlled extraction environments due to highly toxic heavy metals.

da Cunha R C et al. demonstrated a green method which can separate mercury from containing different metals [20]. Although this method is initially for effluents containing mercury, it can also be improved for waste mercury batteries. The study investigated the extraction and separation of Hg(II) using an aqueous two-phase system (ATPS) composed of polyethylene oxide (PEO1500) or triblock copolymers (L64 or L35), an electrolyte (sodium citrate or sodium sulphate), and water, with or without chloride ions. The extraction behaviour of Hg(II) in the phase rich in macromolecules is influenced by various factors, including the amount of added extractant, pH, electrolyte nature, and macromolecule composition of the ATPS. The ATPS consisting of PEO1500 + sodium citrate + H_2O (pH 1.00 and 0.225 mol kg−1 KCl) demonstrated the highest Hg(II) extraction efficiency (%E = 92.3 ± 5.2). Moreover, excellent separation factors (ranging from 1.54×10^2 to 3.21×10^{10}) were achieved for recovering mercury while co-existing metals were present.

Paten CN102117919A also disclosed a method to recover metals from waste mercury batteries [21]. The method involves several steps: removing the negative zinc sheet from the battery, combining the positive substance and electrolyte from the battery with water, stirring to disperse them, adjusting the pH value to a range of 1–6 using acid, thoroughly soaking and stirring to dissolve the electrolyte and mercury compound from the battery into the water, filtering out insoluble substances and recovering them. The filtrate is then reused to soak the positive electrode substance and electrolyte. When the zinc ion concentration in the filtrate reaches 0.1–10 wt%, the negative zinc sheet is placed into the filtrate and stirred to ensure full contact with the filtrate, resulting in the replacement of zinc with mercury ions to form zinc amalgam. Once the mercury ion reaction is complete, the amalgamated zinc sheet is removed from the filtrate and used as an anode, while the filtrate serves as the electrolyte for electrolysis. During electrolysis, the zinc on the anode dissolves, and metallic mercury is separated below the anode and collected, thus completing the mercury recovery process.

12.4 RECYCLING TECHNIQUES OF LITHIUM-ION BATTERIES

Lithium-ion batteries are energy storage devices that utilize lithium ions as a crucial component in their electrochemical processes. Sony Corporation introduced them in 1991, responding to the increasing demand for lightweight rechargeable cells to power the booming market of portable electronic devices [22]. Lithium atoms in the anode become ionized and dissociated from their electrons in discharging. These lithium ions then travel through the electrolyte, migrating toward the cathode. At the cathode, they combine with their electrons, achieving electrical neutrality.

Lithium-ion batteries are one of the most widely used rechargeable battery types, and their dominance continues to grow year after year. Between 2000 and 2010, lithium consumption in batteries increased by 20% annually. In the following decade, that figure jumped to 107% per year for batteries, with overall lithium consumption growing 27% annually on average [23]. Lithium-ion batteries can also be classified based on their electrode materials, with six different types: Lithium Nickel Manganese Cobalt Oxide (NMC), Lithium Nickel Cobalt Aluminum Oxide (NCA), Lithium Iron Phosphate (LFP), Lithium Cobalt Oxide (LCO), Lithium Manganese Oxide (LMO), and Lithium Titanate (LTO) (Table 12.2) [24, 25]. Currently, NMC and LFP are

TABLE 12.2
Comparison of Six Lithium-Ion Batteries (1 Is the Lowest Score, and 5 Is the Highest)

	Energy	Cost	Life Span	Performance	Safety
NMC	5	4	4	4	4
NCA	5	3	4	4	3
LFP	3	4	5	4	5
LCO	5	4	3	4	3
LMO	4	4	3	3	4
LTO	3	2	5	5	5

Source: Miao Yu et al. (2019), BCG, Battery University.

applied extensively in industries. Thus, the following part will introduce their recycling techniques, separately.

12.4.1 Recycling Techniques of Lithium Nickel Manganese Cobalt Oxide (NMC)

Lithium Nickel Manganese Cobalt Oxide (NMC) batteries can be recycled through pyrometallurgy and hydrometallurgy techniques. Pyrometallurgy is a branch of metallurgy that involves using high temperatures to extract and refine metals from various ores and concentrates. It is commonly applied in the recovery of metals from complex mineral ores, including Lithium Nickel Manganese Cobalt Oxide (NMC), which is widely used in the production of lithium-ion batteries. The pyrometallurgical recovery of metals from NMC typically involves the following steps [26, 27]:

1. Battery disassembly: The first step is dismantling the lithium-ion batteries and separating the plastic (Figure 12.1) and the cathode materials containing the NMC compounds. This can be done manually or with the help of automated equipment.
2. Roasting: The NMC cathode materials are subjected to high-temperature roasting in a controlled environment, such as a rotary kiln or a fluidized bed reactor. During this process, the organic binders and other impurities are burned off, leaving behind the metal-rich residue.
3. Smelting: The roasted NMC residue is then mixed with other materials, such as fluxes or reducing agents, and subjected to smelting. Smelting involves heating the mixture to high temperatures in a furnace, where the metals are separated from the slag or other impurities. The molten metals can be tapped and collected, while the slag is discarded.
4. Refining: The collected molten metals may undergo further refining processes to improve their purity and remove any remaining impurities. This can involve techniques such as electrolysis, vacuum distillation, or solvent extraction, depending on the specific metals being recovered.

FIGURE 12.1 Plastic in NMC lithium-ion batteries.

5. Solidification: Once the metals have been refined, they are typically solidified into ingots or other suitable forms for further processing or sale. These metal products can then be used as raw materials in various industries or as feedstock for battery manufacturing.

It's important to note that the pyrometallurgical process for NMC recovery may vary depending on factors such as the specific composition of the NMC material, desired metal recovery rates, and environmental considerations. In addition to pyrometallurgy techniques, hydrometallurgy is commonly applied as well.

Chen W-S et al. focused on enhancing the metal recovery process from lithium-ion batteries (LIBs), specifically lithium nickel manganese cobalt oxide (NMC) cathode waste materials (Figure 12.2), by utilizing hydrometallurgical methods [28]. The acid leaching step was thoroughly investigated to determine the optimal parameters, including acidity concentration, H_2O_2 concentration, leaching time, liquid-solid mass ratio, and reaction temperature, in order to achieve the highest leaching percentage. The cathode material was leached with 2M H_2SO_4 and 10 vol. % H_2O_2 at 70 °C and 300 rpm, using a liquid-solid mass ratio of 30 mL/g. To complete the recovery process, a suitable separation process was designed to retrieve the valuable metals. Cyanex 272 was employed to extract Co and Mn from the leach liquor into the

FIGURE 12.2 Waste cathode materials of NMC.

organic phase. Subsequently, Co and Mn were separated using D2EHPA to obtain high-purity Co. Ni was selectively precipitated with DMG, forming a solid complex. Finally, the remaining Li in the leach liquor was recovered as Li_2CO_3 precipitated by saturated Na_2CO_3, while Co, Mn, and Ni were recovered as hydroxides using NaOH.

Wang W-Y et al. presented a process for efficiently recovering high-purity metallic cobalt from NMC-type Li-ion batteries, which employ lithium nickel manganese cobalt oxide as the cathode material [29]. The study commenced with leaching experiments, where various acid and base solutions were employed to compare the leaching efficiency of cobalt and other metals. To facilitate the recovery process, complete leaching of cobalt was crucial and was successfully achieved through a reductive leaching method. Subsequently, extraction experiments were conducted using different extractants to selectively extract various metals from the leachate solution. The results demonstrated that a sequential and selective separation, with manganese extraction followed by cobalt extraction, yielded the highest cobalt yield and selectivity. P-204 (di(2-ethylhexyl)phosphoric acid) was employed to extract manganese, while P-507 (2-ethyl(hexyl)phosphonic acid mono-2-ethylhexyl ester) was used to extract cobalt. Finally, electrowinning was applied to recover cobalt in its metallic form from the cobalt solution obtained after back-extraction. The overall recovery ratio for cobalt was approximately 93%, with a remarkably high purity of 98.8%.

Xuan W et al. focused on the selective recovery of cobalt(II), nickel(II), and manganese(II) from a leach solution obtained by digesting cathodic material from spent NMC 111 with hydrochloric acid [30]. They employed a series of extraction steps to separate and purify these metals. Initially, cobalt(II) was selectively extracted over manganese(II), using liquid-liquid extraction. This was achieved using a mixture of 0.4 mol L^{-1} Alamine 336 (tri-octyl/decyl amine) diluted in kerosene, which was modified with 10% (vol) 1-dodecanol. The phase volume ratio between the organic phase and the aqueous phase was optimized at O/A = 1/2. Subsequently, manganese(II) was extracted from the cobalt-depleted aqueous phase using 0.7 mol L^{-1} Alamine 336 diluted in kerosene, also modified with 10% (vol) 1-dodecanol, at O/A = 2. Nickel(II) in the leach solution, along with lithium(I), was then precipitated as nickel(II) hydroxide by neutralizing the acidic solution to pH 8 with sodium hydroxide. The recovery process resulted in more than 99.9% cobalt(II) extraction and produced an aqueous phase with 94.6% cobalt(II) and 5.4% manganese(II), which could be subjected to further precipitation steps. The extracted manganese(II) solution had a purity greater than 99.9%. Moreover, the process successfully extracted over 97.0% of nickel(II) from the leach solution, which was subsequently precipitated as nickel(II) hydroxide. The resulting precipitate contained 55.6% nickel(II), 7.85% chloride, 2.42% sodium(I), and 0.02% lithium(I). Finally, the effluent containing 0.649 g L^{-1} lithium(I) and 79.0 g L^{-1} sodium(I) could be treated by solvent extraction using Cyanex 936, a phosphorus-based extractant specifically designed for lithium-sodium separation from chloride media like brines. Alternatively, a precipitation stage could be implemented to selectively recover lithium(I) while leaving sodium(I) behind.

It can be observed that leaching and solvent extraction are the principal procedures for the recovery of metals from waste NMC. The reason is that NMC contains many elements, so the multi-stage extraction process can separate each element with high purity. (Solvent extraction, also known as liquid-liquid extraction or partitioning, is a technique used to separate and purify substances from a mixture based on their differential solubilities in two immiscible liquids. It involves the transfer of a solute (target compound) from one liquid phase (the feed or source phase) into another liquid phase (the solvent or extractant phase). The merits of solvent extraction are high selectivity, automatic operation, and high processing capacity [31].)

12.4.2 Recycling Techniques of Lithium Iron Phosphate

Recycling techniques for lithium iron phosphate (LFP) batteries can be categorized into several methods: pyrometallurgical, hydrometallurgical, bioleaching, and direct recycling [32]. Pyrometallurgical methods involve melting the battery in a high-temperature furnace to recover valuable cathode metals. This approach is most effective for metals like cobalt [33]. Hydrometallurgical recovery techniques for LFP batteries can be further divided into inorganic acid, organic acid, and oxidation reagents. Inorganic acid processes utilize strong acids to dissolve and extract metals from the battery materials, while organic acid methods employ organic acids for the same purpose. Oxidation reagents are used to promote the dissolution of metals through oxidation reactions. Bioleaching, on the other hand, is a promising technique that utilizes microorganisms to extract metals from ores, and it holds potential for recycling LFP batteries. Lastly, direct recycling involves reusing battery components without dismantling them into their individual materials. These methods represent various approaches to recycling LFP batteries and contribute to the sustainable management of battery waste.

Holzer A et al. conducted investigations using a heating microscope to understand the high-temperature behaviour of two cathode materials, LCO and LFP, when combined with carbon in LIBs [34]. The aim was to develop a continuous process for a novel pyrometallurgical recycling method and adapt it to the specific requirements of LIB materials. Two different reactor designs were examined. In the case of LCO, when treated in an Al_2O_3 crucible, approximately 76% of the lithium was effectively removed through the gas stream, making it readily available for further processing. On the other hand, using an MgO crucible resulted in a lithium removal rate of up to 97%. For LFP, the feasibility of the concept for treatment was investigated, and it was observed that 64% of the phosphorus and 68% of the lithium could be effectively removed. These findings provide valuable insights for t developing efficient pyrometallurgical recycling processes for LIB materials.

WU D et al. proposed oxidation pressure leaching to selectively dissolve Li from spent LFP in a stoichiometric sulfuric acid solution [35, 36]. By employing O_2 as an oxidant and sulfuric acid as the leaching agent, over 97% of the Li was successfully leached into the solution. Simultaneously, more than 99% of the iron (Fe) remained in the leaching residue, which allowed for a cost-effective one-step separation of Li and Fe. Subsequently, the pH of the leachate was adjusted, leading to the recovery of over 95% of the Li in the form of LFP product through the removal of iron and a chemical precipitation of phosphate.

In addition to leaching research applying inorganic acids, the literature has reported organic acids leaching due to availability and biocompatibility [35]. Yang Y et al. conducted a study investigating the use of CH_3COOH as a leaching agent for selective Li extraction from spent LFP [37]. The spent batteries were discharged and then immersed in a 5%wt NaCl solution. Subsequently, the batteries were dismantled into their constituent parts, including the cathode, anode, plastics, and metal case. The leaching process was carried out for 30 minutes under specific conditions: 0.8 mol/L CH_3COOH concentration, 6%vol. H_2O_2 addition, 120 g/L solid-to-liquid (S/L) ratio, and a temperature of 50 °C. The results showed that the Li extraction efficiency reached 95.05%. Comparing the Li extraction with the leaching of other metals, the ratio of Li percentage to Fe percentage was found to be 94.8%, indicating selective leaching of Li. Li L et al. experimented on the leaching of LFP using $C_6H_8O_7$ (citric acid) as the leaching agent [38]. In their investigation, the leaching and grinding stages were carried out simultaneously. Various factors were examined to determine their effects on the leaching process, including the mass ratio of $C_6H_8O_7$ to LFP (ranging from 10 to 80), the ball-to-powder ratio (BPR) (ranging from 15 to 55 g/g), the volume of H_2O_2 (ranging from 0 to 2 mL), the grinding time (ranging from 0.5 to 4 hours), and the rotation speed (ranging from 100 to 500 rpm). Through experimental analysis, the researchers identified the optimal conditions, which included, a BPR of 45, a $C_6H_8O_7$ -to-LFP ratio of 20 g/g, an 8-hour grinding time, a rotation speed of 300 rpm, and 1 mL of H_2O_2. Under these conditions, the extraction efficiencies for Li and Fe were determined to be 97.82% and 3.86%, respectively.

On the other hand, bioleaching and direct recycling are also useful methods for treating waste LFP. Bioleaching is a process that utilizes microorganisms to extract valuable metals from ores, minerals, and wastes. It is an environmentally friendly and sustainable alternative to traditional methods. In bioleaching, microorganisms obtain energy from the oxidation process, aiding in the dissolution of metals. The leaching solution, also known as the bioleachate, containing the metal ions can then be further processed to recover the target metals. Direct recycling has been identified as an economically viable method, with minimal adverse environmental impacts and energy-efficient. This approach enables the recovery of all battery components, including anodes, foils, and electrolytes. It is particularly well-suited for LFP type batteries. The direct recycling process demonstrates practical feasibility, making it a promising solution for sustainable battery waste management. Nevertheless, direct recycling does have several drawbacks that need to be addressed. Firstly, the process is currently at a low maturity level, primarily limited to laboratory-scale experiments. Therefore, it will require significant time and effort to advance and reach commercialization. Secondly, the development of a regeneration process is still in progress, which is crucial for ensuring the longevity and sustainability of the recycling approach. Lastly, the mixing of different cathode materials during the recycling process can result in decreased performance and reduced value of the recycled products. These challenges highlight the need for further research and development to overcome these limitations and optimize the direct recycling of batteries [39–41].

12.5 RECYCLING TECHNIQUES OF LEAD-ACID BATTERIES

A lead-acid battery is a type of rechargeable battery that uses lead and sulfuric acid to function. It was first invented in 1859 by French physicist Gaston Planté. Here is a brief introduction to lead-acid batteries [6]:

Composition: A lead-acid battery consists of a negative electrode made of spongy or porous lead and a positive electrode made of lead oxide. Both electrodes are immersed in an electrolytic solution of sulfuric acid and water.

Operation: When the battery is discharged, a chemical reaction occurs between the lead and lead oxide electrodes and the sulfuric acid electrolyte, producing electricity. This reaction is reversed during the charging process, allowing the battery to be recharged.

Advantages: Lead-acid batteries have a well-established and mature technology base, making them the most commonly applied battery for many applications, such as starting car engines. They also have the ability to supply high surge currents, giving them a relatively large power-to-weight ratio.

Applications: Lead-acid batteries are widely used in various applications, including backup power supplies in cell phone towers, high-availability emergency power systems like hospitals, and stand-alone power systems. They are also commonly used in automotive applications. Despite their relatively low energy density compared to newer battery technologies, lead-acid batteries remain popular due to their low cost and ability to deliver large surges of electricity.

Due to the various application, the waste number of lead-acid batteries increased and needed to be treated. Lead-acid batteries contain lead, which is a toxic heavy metal. Improper disposal of lead-acid batteries can lead to environmental contamination and pose risks to human health. However, lead-acid batteries are also one of the most recycled consumer products, with a high recycling rate due to the value of the lead and other materials they contain.

Recycling used lead-acid batteries is crucial as it prevents them from entering the waste stream and being disposed of in landfills, where the lead can contaminate groundwater. By recycling, we can avoid the release of lead into the environment and eliminate the energy-intensive process of manufacturing lead from new resources. The economic viability of obtaining secondary lead from used batteries depends on the market price of lead, but it generally requires less energy compared to producing primary lead from ore. Furthermore, recycling reduces the dispersion of lead in the environment and preserves mineral resources for future generations when done in an environmentally and socially responsible manner. Here are some recycling techniques and experience with lead-acid batteries [42]:

1. Crushing: Lead-acid batteries are crushed during recycling into pieces that measure about the size of a nickel. The various components are separated out of the pieces. Crushing is done inside a hammer mill before the broken parts are inserted into a vat. When inside the vat, the heavy materials and lead drop to the bottom and the plastic floats.
2. Smelting: Smelting is another technique used for recycling lead-acid batteries. The process involves melting the lead and separating it from other materials. The lead is then used to make new batteries or other products.
3. Hydrometallurgical process: This process involves the use of chemicals to dissolve the lead from the battery. The lead is then recovered from the solution using various techniques. This process is more expensive than smelting but can recover more lead.
4. Closed-loop recycling: Lead-acid batteries are closed-loop recycled, which means each part of a battery is recycled into a new battery. According to the EPA (Environmental Protection Authority), about 80% of the lead and plastic in a lead-acid battery is recycled for reuse.

Ferracin L C et al. recovered lead from waste lead-acid batteries and turned it into raw materials for perovskite solar cells [43]. In their research, they successfully extracted lead from lead-acid batteries by reacting the anode and cathode lead mud with acetic acid. This process produced high purity lead acetate ($Pb(Ac)_2$), which was characterized using FTIR and XRD techniques. The synthetic path employed is not only efficient but also environmentally friendly, as it avoids causing secondary pollution. The recovered lead acetate was then utilized to fabricate normal planar heterojunction perovskite solar cells (PerSCs), achieving an impressive power conversion efficiency of 17.83%. Interestingly, besides lead acetate, they also found CH_3COOH in the product obtained from the cathode mud. This compound is beneficial for obtaining a compact and crystalline perovskite film, ultimately enhancing the performance of solar cells. By utilizing lead from spent batteries in the fabrication of perovskite solar cells, it can contribute to reducing the environmental impact of battery waste while advancing the development of renewable energy technology.

Buzatu et al. proposed a method that involves leaching with NaOH, which is not only useful for processing secondary resources like volatile dusts resulting from lead extraction from primary sources but also applicable to the treatment of oxidic paste obtained from dismantled lead-acid batteries [44]. Through chemical analysis and X-ray diffraction, it has been confirmed that the oxidic paste from spent batteries contains approximately 70–73% Pb, predominantly as anglesite (around 38% Pb_SO_4) and lanarkite (approximately 36% Pb_2SO_5). The researchers conducted experiments involving NaOH leaching of both the industrial paste obtained from spent car batteries and pure substances. The experiments were carried out at temperatures of 40 °C, 60 °C, and 80 °C, with varying solid/liquid ratios (S:L = 1:10, 1:30, and 1:50), and using different molarities of NaOH solutions (2M, 4M, 6M, and 8M). The leaching time was 1 hour for the pure substances and 2 hours for the industrial pastes. By employing a 6M NaOH leaching solution at 60 °C for 2 hours, with S:L ratios of 1:20–1:30, an extraction efficiency of 92% Pb was achieved. However, the efficiency could be increased up to 97% Pb only when the leaching process was conducted in two steps.

Li F et al. introduced a mechanism for the efficient recovery of lead and iron by identifying spatial position relationships [45]. The results revealed the presence of wrapping, embedding, and loading spatial position relationships between PbS and Fe_2SiO_4. Upon the disruption of Fe_2SiO_4, PbS became exposed on the surface of DR-LABs (dismantled lead-acid batteries). Subsequently, through acid leaching followed by pH adjustment, 98.9 wt% of lead was successfully recovered as $Pb(OH)_2/Fe(OH)_3/Fe(OH)_2$. Simultaneously, 90.3% of iron was recovered as raw material. Experimental analysis and DFT (Density Functional Theory) calculation confirmed that NaOH played a vital role in inducing the formation of hydroxylated $Fe_2SiO_4(010)$ surfaces and facilitating near-surface Na+ substitution. These factors promoted the transformation of Fe_2SiO_4 to FeO. Overall, this proposed mechanism holds significant promise for the efficient recovery of lead and iron from DR-LABs.

Based on the abovementioned situations and recycling techniques, it can be understood that battery recycling plays a crucial role in environmental protection, resource conservation, energy efficiency, regulatory compliance, and safeguarding public health. It is essential to raise awareness, promote recycling programs, and ensure proper collection and recycling infrastructure to maximize the benefits of battery recycling.

REFERENCES

1. Dell R M 2007 *Understanding Batteries.* Royal Society of Chemistry.
2. Pistoia G 2005 *Batteries for Portable Devices.* Elsevier.
3. Rarotra S, Sahu S, Kumar P, Kim K, Tsang YF, Kumar V, et al. 2020 Progress and challenges on battery waste management: A critical review *ChemistrySelect* **5** 6182–93.
4. Britain G 1996 *State Audit in the European Union.* Commercial Colour Press.
5. Dobrowolski Z, Sułkowski Ł and Danielak W 2021 Management of waste batteries and accumulators: Quest of European Union goals *Energies* **14** 6273.

6. Garche J 2013 *Encyclopedia of Electrochemical Power Sources*. Newnes.
7. Prabhansu 2022 *Emerging Trends in Energy Storage Systems and Industrial Applications*. Elsevier
8. Ferella F, De Michelis I and Vegliò F 2008 Process for the recycling of alkaline and zinc–carbon spent batteries. *J. Power Sources* **183** 805–11.
9. Shin S M, Senanayake G, Sohn J, Kang J, Yang D and Kim T 2009 Separation of zinc from spent zinc-carbon batteries by selective leaching with sodium hydroxide *Hydrometall.* **96** 349–53.
10. Khan M H, Gulshan F and Kurny A S W 2013 Recovery of metal values from spent zinc–carbon dry cell batteries *J. Inst. Eng. India. Ser. D* **94** 51–6.
11. Shangguan E, Wang L, Wang Y, Li L, Chen M, Qi J, et al. 2022 Recycling of zinc–carbon batteries into MnO/ZnO/C to fabricate sustainable cathodes for rechargeable zinc-ion batteries *ChemSusChem* **15**. e202200720, 1-9
12. 陳家鏞 2005 濕法冶金手冊. 冶金工業出版社.
13. Owens B B 2012 *Batteries for Implantable Biomedical Devices*. Springer.
14. Reddy T B 2001 *Handbook of Batteries*. McGraw-Hill Professional.
15. Saghazadeh A and Rezaei N 2017 Systematic review and meta-analysis links autism and toxic metals and highlights the impact of country development status: Higher blood and erythrocyte levels for mercury and lead, and higher hair antimony, cadmium, lead, and mercury *Prog. Neuro-Psychopharmacol. Biol. Psychiatry* **79** 340–68.
16. Tsai W-T 2022 Multimedia pollution prevention of mercury-containing waste and articles: Case study in Taiwan *Sustainability* **14** 1557.
17. Chen W-S, Chi C-C and Lee C-H 2023 Stabilization of waste mercury with sulfide through the Ball-Mill method and heat treatment *Sustainability* **15** 10333.
18. He K, Xun P, Liu K, Morris S, Reis J and Guallar E 2013 Mercury exposure in young adulthood and incidence of diabetes later in life *Diabetes Care* **36** 1584–9.
19. Tsai T-L, Kuo C-C, Pan W-H, Wu T-N, Lin P and Wang S-L 2019 Type 2 diabetes occurrence and mercury exposure – From the National Nutrition and Health Survey in Taiwan *Environ. Int.* **126** 260–7.
20. da Cunha R C, Patrício P R, Vargas S J R, da Silva L H M and da Silva M C H 2016 Green recovery of mercury from domestic and industrial waste *J. Hazard. Mater.* **304** 417–24.
21. 黄启明, 李伟善, 莫烨强, 罗建成, 谭春林, 吕东升CN102117919A - Method for recovering mercury from waste neutral zinc-manganese dioxide battery - Google Patents Available from: https://patents.google.com/patent/CN102117919A/en
22. Qiao H, Wei Q 2012 *Functional Nanofibers in Lithium-Ion Batteries* Woodhead Publishing Limited. 197–208
23. Govind Bhutada Lithium consumption has nearly quadrupled Since 2010 Available from: https://elements.visualcapitalist.com/lithium-consumption-has-nearly-quadrupled-since-2010/
24. van Schalkwijk W and Scrosati B 2002 *Advances in Lithium Ion Batteries Introduction* Springer New York, NY 1–5.
25. Pistoia G 2013 *Lithium-Ion Batteries*. Newnes.
26. dos Santos M P, Garde I A A, Ronchini C M B, Filho L C, de Souza G B M, Abbade M L F, et al. 2021 A technology for recycling lithium-ion batteries promoting the circular economy: The RecycLib. *Resour. Conserv. Recycl.* **175** 105863.
27. Ciez R E, Whitacre J F 2019 Examining different recycling processes for lithium-ion batteries *Nat. Sustain.* **2** 148–56.
28. Chen W-S and Ho H-J 2018 Recovery of valuable metals from lithium-ion batteries nmc cathode waste materials by hydrometallurgical methods *Metals* **8** 321.

29. Wang W-Y, Yen CH, Lin J-L and Xu R-B 2019 Recovery of high-purity metallic cobalt from lithium nickel manganese cobalt oxide (NMC)-type Li-ion battery *J. Mater. Cycles Waste. Manag.* **21** 300–7.
30. Xuan W, de Souza Braga A and Chagnes A 2022 Development of a novel solvent extraction process to recover cobalt, nickel, manganese, and lithium from cathodic materials of spent lithium-ion batteries *ACS Sustain. Chem. Eng.* **10** 582–93.
31. Rydberg J 2004 *Solvent Extraction Principles and Practice, Revised and Expanded.* CRC Press.
32. Larouche F, Tedjar F, Amouzegar K, Houlachi G, Bouchard P, Demopoulos G P, et al. 2020 Progress and status of hydrometallurgical and direct recycling of li-ion batteries and beyond *Materials* **13** 801.
33. Baum Z J, Bird R E, Yu X, Ma J 2022 Lithium-ion battery recycling - Overview of techniques and trends *ACS Energy Lett.* **7** 712–9.
34. Holzer A, Windisch-Kern S, Ponak C, Raupenstrauch H 2021 A novel pyrometallurgical recycling process for lithium-ion batteries and its application to the recycling of LCO and LF *Metals* **11** 149.
35. da Vasconcelos D S, Tenório J A S, Botelho Junior A B, Espinosa D C R 2023 Circular recycling strategies for LFP batteries: A review focusing on hydrometallurgy sustainable processing*Metals* **13** 543.
36. Wu D, Wang D, Liu Z, Rao S and Zhang K 2022 Selective recovery of lithium from spent lithium iron phosphate batteries using oxidation pressure sulfuric acid leaching system *Trans. Nonferrous Met. Soc.* **32** 2071–9.
37. Yang Y, Meng X, Cao H, Lin X, Liu C, Sun Y, et al. 2018 Selective recovery of lithium from spent lithium iron phosphate batteries: a sustainable process *Green Chem.* **20** 3121–33.
38. Li L, Bian Y, Zhang X, Yao Y, Xue Q, Fan E, et al. 2019 A green and effective room-temperature recycling process of $LiFePO_4$ cathode materials for lithium-ion batteries *Waste Manag.* **85** 437–44.
39. Ali H, Khan H A and Pecht M 2022 Preprocessing of spent lithium-ion batteries for recycling: Need, methods, and trends *Renew. Sust. Energ. Rev.* **168** 112809.
40. Roy J J, Rarotra S, Krikstolaityte V, Zhuoran K W, Cindy Y D, Tan X Y, et al. 2022 Green recycling methods to treat lithium-ion batteries E-waste: A circular approach to sustainability *Adv. Mater.* **34** 2103346.
41. Biswal B K and Balasubramanian R 2023 Recovery of valuable metals from spent lithium-ion batteries using microbial agents for bioleaching: a review *Front. Microbiol.* **14**. 1197081-1197105
42. Ogundele D, Ogundiran M B, Babayemi J O and Jha M K 2020 Material and substance flow analysis of used lead acid batteries in Nigeria: Implications for recovery and environmental quality *J. Health Pollut.* **10** 200913.
43. Ferracin L C, Chácon-Sanhueza A E, Davoglio R A, Rocha L O, Caffeu D J, Fontanetti A R, et al. 2002 Lead recovery from a typical Brazilian sludge of exhausted lead-acid batteries using an electrohydrometallurgical process *Hydrometall.* **65** 137–44.
44. Buzatu T, Petrescu M I, Ghica V G, Buzatu M and Iacob G 2015 Processing oxidic waste of lead-acid batteries in order to recover lead *Asia Pac. J. Chem. Eng.* **10** 125–32.
45. Li F, Wei X, Chen Y, Zhu N, Zhao Y, Cui B, et al. 2022 Efficient recovery of lead and iron from disposal residues of spent lead-acid batteries *Resour. Conserv. Recycl.* **187** 106614.

13 Recycling Solar Cell Materials

Wei-Sheng Chen and Cheng-Han Lee
National Cheng Kung University, Tainan City, Taiwan

13.1 BACKGROUND

13.1.1 Introduction of Solar Cells

Solar cells, also known as photovoltaic (PV) cells, are devices that convert sunlight directly into electricity, using the photovoltaic effect [1]. They are a vital component of solar panels and play a key role in harnessing solar energy for various applications [1]. The concept of solar cells dates back to the 19th century, when French physicist Alexandre-Edmond Becquerel discovered the photovoltaic effect in 1839. The photovoltaic effect is the phenomenon where certain materials generate an electric current when exposed to light. This discovery laid the foundation for the development of solar cell technology.

Solar cells are typically made from semiconducting materials, most commonly crystalline silicon [2]. Silicon is abundantly available in the Earth's crust and possesses the necessary properties to efficiently convert sunlight into electricity. Other semiconductor materials, such as cadmium telluride and copper indium gallium selenide, are also applied to produce solar cells. The basic structure of a solar cell consists of several layers. The top layer is a transparent protective cover that allows sunlight to pass through. Below this layer is a thin, negatively charged layer called the N-type layer. It contains atoms with extra electrons. The bottom layer is a positively charged layer called the P-type layer, which has atoms with missing electrons, creating "holes" in the material. These layers together form a semiconductor junction known as a P-N junction [3].

When sunlight strikes the solar cell, photons (particles of light) transfer their energy to the semiconductor material. This energy excites the electrons in the N-type layer, allowing them to break free from their atoms and move toward the P-type layer [4]. The movement of these electrons creates an electric current, which can be harnessed for various applications. To collect the generated electricity, metal contacts are attached to the top and bottom layers of the solar cell. These contacts allow the current to flow out of the cell and be used to power electrical devices or stored in batteries for later use.

Solar cells can be connected in series or parallel to form solar panels. Multiple solar panels can be installed together to create larger solar arrays, which are commonly used in solar power systems for residential, commercial, and industrial applications. The electricity generated by these solar arrays can be used to power homes,

 DOI: 10.1201/9781003463009-13

businesses, and even contribute to the grid, reducing reliance on fossil fuels and mitigating environmental impacts [5].

Over the years, advancements in solar cell technology have improved their efficiency, durability, and affordability. Research and development efforts focus on enhancing solar cell performance and exploring new materials and manufacturing techniques. Solar cells have become a promising and increasingly important renewable energy technology, contributing to a more sustainable and cleaner future.

13.1.2 Types of Solar Cells

There are several types of solar cells available, including Amorphous Silicon solar cells (a-Si), Biohybrid solar cells, Cadmium telluride solar cells (CdTe), Concentrated PV cells (CVP and HCVP), Copper indium gallium selenide solar cells (CIGS), Crystalline silicon solar cells (c-Si), Float-zone silicon, Dye-sensitized solar cells (DSSC), Gallium arsenide germanium solar cells (GaAs), Hybrid solar cell, Luminescent solar concentrator cells (LSC), Micromorph (tandem-cell using a-Si/μc-Si), Monocrystalline solar cells (mono-Si), Multi-junction solar cells (MJ), Nanocrystal solar cells, Organic solar cells (OPV), Perovskite solar cells, Photoelectrochemical cells (PEC), Plasmonic solar cells, Polycrystalline solar cells (multi-Si), Quantum dot solar cells, Solid-state solar cells, Thin-film solar cells (TFSC), and Wafer solar cells (wafer-based solar cell crystalline).

Solar cells can be categorized into three generations: first, second, and third. First-generation solar cells are made from highly pure silicon and have a single junction to extract energy from photons. They are known for their high efficiency, approaching the theoretical maximum of 33%. These cells remain the most efficient option for residential use. Second-generation solar cells, such as amorphous silicon, CdTe, and CIGS cells, are thin-film solar cells that are widely used in utility-scale photovoltaic power stations and buildings. Third-generation solar cells are still in the research phase. The aim of third-generation solar energy research is to develop cost-effective, high-efficiency cells that utilize innovative approaches to achieve efficiencies ranging from 30% to 60% [6–10]. The introduction of common solar cells is shown below and Table 13.1 [11, 12].

TABLE 13.1
List of Different Solar Cells [24]

First Generation	Efficiency	Second Generation	Efficiency	Third Generation	Efficiency
Single Crystal	26.8%	CIGS	23.4%	Dye-Sensitized Solar Cells	11.9%
Multicrystalline	23.1%	CdTe	21.0%	Perovskite Solar Cells	25.8%
Thin Film Crystal	23.3%			Multi-Junction Solar Cells	38.8%

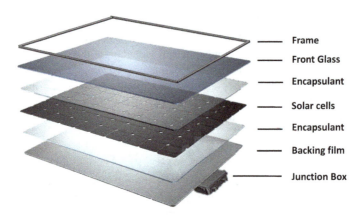

FIGURE 13.1 Structure of solar panel [13].

1. Crystalline Silicon Cells: This is the most common type of solar cells made from a single crystal of silicon. They are highly efficient and long-lasting. The structure of typical crystalline silicon cells is revealed above (Figure 13.1) [13].
2. Polycrystalline Cells: Polycrystalline cells, also known as multicrystalline cells, are a type of solar cell used to convert sunlight into electricity. Unlike first-generation solar cells that use extremely pure silicon, polycrystalline cells are made from silicon wafers with multiple crystal structures. These crystals are less uniform than those found in monocrystalline cells, resulting in a characteristic "grainy" appearance. They are more cost-effective than monocrystalline cells but are slightly less efficient.
3. Thin-Film Solar Cells: Unlike traditional solar cells, typically made from silicon wafers, thin-film solar cells are made by depositing one or more thin layers of PV material on a supporting material such as glass, plastic, or metals. There are several types of thin-film solar cells, including cadmium telluride (CdTe) and copper indium gallium diselenide (CIGS). They are less efficient than crystalline silicon cells but are also less expensive [14, 15].
4. Biohybrid Solar Cells: Biohybrid solar cells are a type of solar cell that combines the principles of biological systems with traditional photovoltaic technology. These cells aim to leverage the unique properties of biological materials, such as enzymes, proteins, or other biomolecules, to enhance the efficiency and functionality of solar energy conversion. They are still in the experimental stage and not yet commercially available [16, 17].
5. Concentrated PV Cells: These cells are a specialized type of solar cell that utilizes optical systems to concentrate sunlight onto small, high-efficiency photovoltaic cells. CPV technology focuses sunlight using lenses or mirrors onto the cells, significantly increasing the intensity of sunlight that reaches them. By concentrating sunlight, CPV cells can achieve higher conversion efficiencies than traditional flat-plate solar cells.
6. Dye-Sensitized Solar Cells: These cells employ a layer of photosensitive dye, typically a specialized organic dye or a metal-organic complex, coated

onto a porous semiconductor material such as titanium dioxide. The dye absorbs sunlight, exciting its electrons and generating an electric current. The excited electrons are transported through the semiconductor material and collected at the electrodes, generating electricity [18, 19].

7. Perovskite solar cells: Perovskite is a mineral with the ability to absorb light and utilize a minimal amount of material to capture sunlight compared to other solar cells. It is a semiconductor that can transport an electric charge when exposed to light. It has gained attention for its high efficiency and potential for low-cost production. However, there are still challenges to overcome before they can become a competitive commercial technology [20, 21].
8. Luminescent Solar Concentrator Cells: These cells consist of a transparent plate or waveguide coated with a luminescent dye or semiconductor nanoparticles. When sunlight strikes the plate, it is absorbed by the luminescent material, which then emits light at longer wavelengths. This emitted light is trapped within the plate due to total internal reflection and travels to the edges, where small solar cells are placed to convert the trapped light into electricity [22, 23].
9. Micromorph Cells: Micromorph cells are thin-film solar cells that combine amorphous silicon (a-Si) and microcrystalline silicon (μc-Si) layers in a tandem structure. This tandem configuration allows for enhanced light absorption and improved efficiency compared to single-junction thin-film solar cells.
10. Multi-Junction Solar Cells: These types of solar cells use multiple layers of different materials to capture different wavelengths of sunlight. They are highly efficient but expensive and are mostly employed in space applications.

13.1.3 Situation of Waste Solar Cells

Waste solar cells, end-of-life solar cells, or photovoltaic (PV) waste refer to solar panels that have reached the end of their useful life or have been damaged and are no longer functioning properly. As with any electronic waste, the disposal of solar cells requires careful consideration to minimize environmental impact and maximize resource recovery. The exact amount of waste solar cells globally is challenging to determine due to several factors, such as the varying lifespan of solar panels, the growth of the solar industry, and the need for comprehensive data on the disposal and recycling of solar panels. However, according to the report, approximately 80 million tons of PV waste will accumulate in 2050 [25] (Take Taiwan, for instance. The amount of solar panel waste is projected to reach 10,000 tons after 2023 and 100,000 tons by 2035). Currently, the situation of waste solar cells is as follows:

1. Growing Waste: The solar industry has been experiencing rapid growth over the years, resulting in an increasing number of solar panels installed worldwide. With the average lifespan of solar panels ranging from 25 to 30 years, the number of discarded or end-of-life solar panels is expected to continue to rise in the coming decades.

2. Recycling Challenges: While solar panels are designed to be durable and long-lasting, they eventually reach the end of their life cycle. The recycling of solar panels is complex due to the presence of various materials, including glass, aluminium, silicon cells, and rare metals such as silver, indium, and gallium. These materials require specialized processes for effective separation and recycling, which can be expensive and technically challenging [26].
3. Limited Recycling Infrastructure: Currently, the infrastructure for recycling solar panels is not as well-established as it is for other recyclable materials. Many countries need more specific regulations or guidelines for properly disposing of and recycling solar panels. Consequently, a significant proportion of end-of-life solar panels are often landfilled or incinerated, leading to environmental concerns.
4. Emerging Recycling Technologies: Despite the challenges, researchers and companies are actively working on developing innovative recycling technologies for solar panels. These technologies aim to recover valuable materials, minimize environmental impact, and reduce the cost of recycling. For instance, mechanical shredding, thermal treatment, and chemical extraction are being explored to recover materials from decommissioned solar panels efficiently [27].
5. Extended Producer Responsibility (EPR): Some regions have started implementing Extended Producer Responsibility (EPR) regulations, which hold solar panel manufacturers accountable for the end-of-life management of their products. EPR encourages manufacturers to design products with recycling in mind and establish collection and recycling networks to handle the waste generated.

Overall, the situation of waste solar cells highlights the need for improved recycling practices and regulations to address the growing volume of end-of-life panels and prevent environmental harm. The following paragraph will introduce the recycling steps of solar cell materials and the recycling techniques of some different types of solar cells.

13.2 RECYCLING TECHNIQUES OF SOLAR CELL MATERIALS

Recycling techniques for solar cell materials involve several steps and methods. The main point of recycling solar cells is delamination and material separation. PV recycling typically involves three main stages: delamination, material separation, and material recovery. Delamination refers to the separation of different layers of the solar panel, while material separation involves separating the different components of the panel. Material recovery focuses on recovering valuable materials for reuse. Here is the information on each step mentioned in other research [28–33].

1. Removal of Frame and Junction Box: To prepare the panels for recycling, the frame and junction box are typically removed from the solar cells. Dismantling processes may involve manual or automated methods, depending on the recycling facility's capabilities and the volume of processed

panels. This step aims to separate the different components of the solar panels for further recycling.
2. Removal of Glass: The glass layer is typically removed from the solar cells using specialized equipment or processes. One common method is mechanical separation, where the panels are crushed or shredded, allowing the glass to be separated from the other layers, such as the semiconductor materials and backsheet.
3. Recycling of Glass, Aluminium Frame, and Plastic Junction Box: Many components of solar panels, such as glass, the aluminium frame, and the plastic junction box, are easily recyclable. These materials can be processed through established recycling industries. For example, the frames of solar panels are often made of aluminium, a valuable and recyclable material. The removed frames are typically sent for aluminium recycling, where they are processed to recover the metal for reuse in various industries. The plastic components in the junction box can be sent for plastic recycling or energy recovery, like incineration with energy capture.
4. Separation and Purification of Metals: After removing the frame, junction box, and glass, the solar cells themselves can undergo further recycling. The processing may involve mechanical shredding or thermal treatment to separate the different layers and recover valuable materials, such as silicon, silver, copper, indium, and other metals, using chemical and electrical techniques. This allows for the recovery of valuable materials from the solar panels.

13.2.1 Recycling Techniques of Crystalline-Silicon (c-Si) Solar Cell (First Generation)

The recycling of crystalline silicon (c-Si) solar cells typically involves a combination of mechanical separation, thermal treatment, chemical etching, and various purification processes. Mechanical separation is applied to break down the cells into smaller components, while thermal treatment helps remove organic materials such as ethylene vinyl acetate (EVA). Chemical etching is employed to dissolve the antireflective coating from the silicon wafers, which are then reclaimed and cleaned for potential reuse. This process can be classified into using hydrofluoric acid (HF) [34–36] or not using HF [37–39] (HF can dissolve silicon). Metal recovery techniques are employed to extract valuable metals used in the cells, such as silver and copper. Additionally, the glass components are recycled separately by crushing and melting them for other industries. Among the above recycling techniques, chemical etching can obtain silicon, whereas metal recovery techniques can acquire metals such as copper, tin, and lead from PV ribbon.

PV ribbons contain most of the metal in the PV module, and it looks like the wire that carries the electrical energy produced by solar cells. Lee J-S et al. conducted a study to purify copper from spent photovoltaic ribbon by utilizing a zone-melting furnace, producing high-purity copper with a grade of 4N [40]. The initial step involved oxidation at temperatures ranging from 300 to 900 °C to eliminate the coating layer. The oxidized coating layer, composed of 68.99 wt.% lead and 31.21 wt.% tin, was subsequently removed from the substrate at room temperature. The chemical composition of the copper ribbon analyzed through ICP-MS after oxidation revealed a purity of approximately 99.5 wt.%. Following this, a further process utilizing a zone-melting furnace was conducted, resulting in the attainment of copper with a grade of 4N (≥99.99%).

Moon G et al. proposed a recycling method involving HCl leaching with Sn^{4+} followed by solvent extraction has been proposed to recover copper (Cu), tin (Sn), and lead (Pb) from photovoltaic (PV) ribbon [41]. During the HCl leaching process with Sn^{4+}, Sn and Pb were separated from the PV ribbon, while Cu remained as a plate. The leaching efficiency of Sn improved with higher temperatures, increased initial Sn^{4+} concentration, or decreased pulp density. In a leaching test conducted at 400 rpm and 70 °C, using a 1 M HCl solution with 5% pulp density and 5000 mg/L Sn^{4+}, the leaching efficiency of Sn reached over 99% within 60 minutes. Subsequently, 71.9% of Pb was recovered as $PbCl_2$ powder after allowing the leach solution to settle for 24 hours at room temperature. Selective extraction of Sn from the leach solution was achieved through solvent extraction, using tri-butyl phosphate (TBP). The efficiency of Sn extraction increased with higher TBP content in kerosene, while less than 1% of Pb was extracted. Consequently, the successful separation of Sn, Cu, and Pb was accomplished through the recycling process.

Chen W-S et al. provided an effective recycling approach for valuable materials obtained from waste crystalline-silicon PV modules, encompassing Si within the PV cell, as well as Ag, Cu, Pb, and Sn within the PV ribbon [13, 42]. After removing the tempered glass and Ethylene Vinyl Acetate (EVA) resin, the module was divided into PV cells and PV ribbons (Figure 13.2). The purification process for the PV cell involved the recovery of Si with a purity of 99.84% through a two-step leaching process that eliminated impurities such as aluminium and silver by dissolving them (Five moles of nitric acid and one mole of potassium hydroxide were employed to

FIGURE 13.2 PV cells and PV ribbons.

silicon purification two-step leaching. 99.7% of silver and 98.9% of aluminium were dissolved and separated). To recover the PV ribbon, different techniques were proposed during the metal separation process. Hydrochloric acid was utilized for leaching the PV ribbon, effectively extracting the desired metals. The separation of silver from other metals was achieved through halogenation, resulting in the formation of AgCl. After pretreatment, metals, including Cu, Pb, and Sn, were then ionized into liquid. Both TBP and LIX984N were diluted in kerosene and employed as extractants for separating Cu, Pb, and Sn. The extraction process consisted of two steps. In the first step, TBP was utilized to extract Sn from the liquid, followed by stripping it using nitric acid. In the second step, LIX984N was employed to extract Cu from the liquid, subsequently stripping it with sulfuric acid. After all the processes were finished, the purities of final products in this research are 98.85% of silver, 99.7% of CuO, 99.47% of PbO, and 99.68% of SnO_2, and the recovery rates of each are 98.5%, 96.5%, and 88.9%, respectively.

Sah D et al. investigated a recovery process that successfully obtained materials such as glass, junction box, polymer back sheet, and aluminium frame without causing any damage. The EVA layer was separated by treating the panel with a toluene solution. An optimized solution of 2 M HCl was utilized to strip Pb and Sn from the connecting wires, enabling the recovery of Cu strips. Pb was obtained as $PbCl_2$ by precipitating it from the solution using an ice bath, while Sn was recovered as SnO_2 and $SnCl_2$ through electrolysis of the residual solution. The Al from the back contact was recovered as $Al(OH)_3$ by sequentially treating it with KOH and H_2SO_4 solutions. The impact of the molar concentration of the KOH solution on Al etching was also investigated, revealing that the 2 M solution yielded optimal results. A cost-effective approach was implemented to recover silver (Ag), which involved treating solar cell pieces with HNO_3 acid, followed by the reduction of $AgNO_3$ using copper strips. Reusing recovered Cu strips for Ag recovery served as a cost-reducing step in recycling. At last, the antireflective coatings were removed using an H_3PO_4 solution, resulting in the retrieval of silicon wafer pieces.

Based on the abovementioned literature, it can be found that the recycling techniques of crystalline-silicon (c-Si) solar cell is thriving. Different steps and combinations can be employed to obtain targeted materials. In the future, people can recover materials through distinct methods, depending on the need and cost.

13.2.2 Recycling Techniques of CIGS Solar Cells (Second Generation)

Copper indium gallium diselenide (CIGS) solar cells are thin-film solar cells that convert sunlight into electricity. Like other electronic devices, these cells have a limited lifespan and may eventually need to be recycled or disposed of. Recycling CIGS solar cells is vital in recovering valuable materials and minimizing environmental harm. While CIGS solar cell recycling is still evolving, various techniques are currently being researched and investigated.

1. Mechanical Separation: This technique involves the physical dismantling of CIGS solar cells to separate the components. The cells are crushed or shredded, and the resulting mixture is subjected to various mechanical separation

processes, such as sieving and gravity separation, to separate the different layers and materials.

2. Thermal Treatment: Thermal treatment techniques involve subjecting CIGS solar cells to high temperatures to break down the materials into their constituent components. This can include pyrolysis or incineration. The heat causes the decomposition of organic materials and the vaporization or sublimation of some volatile compounds, leaving behind residues that can be further processed.
3. Chemical Leaching: Chemical leaching involves applying various chemical solutions to dissolve or extract specific materials from the CIGS solar cells. Different chemicals are used to selectively dissolve and separate the different layers, such as the CIGS absorber layer and the transparent conductive oxide layer. This allows for the recovery of valuable materials like indium, gallium, and selenium [43].
4. Metal Recovery: Metal recovery can be conducted through electrochemical methods and hydrometallurgical processes. Electrochemical methods use electrolysis or other electrochemical reactions to dissolve or extract materials from CIGS solar cells. These methods typically involve immersing the cells in an electrolyte solution and applying an electric current to induce the desired reactions. Hydrometallurgical processes involve using aqueous solutions and chemical reactions to extract valuable metals from CIGS solar cells. These processes can involve leaching, precipitation, solvent extraction, ionic liquid extraction ion-exchange, and other chemical techniques to separate and recover the different materials present in the cells [44].

Gustafsson A M K et al. investigated the possibility of recycling selenium from CIGS solar cells by subjecting them to oxidation at high temperatures [45]. This process led to the formation of gaseous selenium dioxide, which could be separated from the other elements that remained in a solid state. The selenium dioxide sublimed upon cooling and could be collected as a crystalline material. By oxidizing the CIGS material at 800 °C for one hour, selenium was successfully separated. Two different reduction methods were tested to convert the selenium dioxide into selenium. In the first method, an organic molecule was used as a reducing agent in a Riley reaction. In the second method, sulphur dioxide gas was employed. Both approaches yielded high-purity selenium. These findings demonstrate that the proposed method for selenium separation could serve as the initial step in a recycling process to achieve complete separation and recovery of high-purity elements from CIGS solar cells.

Zimmermann Y-S et al. conducted research in which various nanofiltration membranes were evaluated for their effectiveness in extracting indium from leachates of CIGS photovoltaic cells [46]. The extraction process was carried out under low pH conditions and low transmembrane pressure differences (less than 3 bar). The nanofiltration process retained over 98% of indium, even at highly acidic pH levels, effectively separating it from other elements, such as Ag, Sb, Se, and Zn. Subsequently, a selective liquid-liquid extraction (LLE) technique using di-(2-ethylhexyl)phosphoric acid (D2EHPA) was employed on the retentates. This LLE method extracted 97% of the indium, effectively separating it from all other elements except Mo, Al, and Sn. Approximately 95% (2.4 g m^{-2} CIGS) of the indium could be successfully recovered into the D2EHPA phase. Additionally, the nanofiltration process resulted in a significant reduction (>60%) in the consumption of D2EHPA, thanks to the concentration of metals in the reduced retentate volume. These findings highlight the promising potential for efficiently recovering scarce metals from secondary resources. Furthermore, the applicability of nanofiltration at low pH (≥0.6) makes it suitable for use in hydrometallurgical processes that typically employ acidic conditions.

Liu F-W et al. proposed a separation process for copper (Cu), indium (In), gallium (Ga), and selenium (Se) present in CIGS-based thin-film solar panels [47]. The process begins with layer-by-layer peeling of the panels, taking advantage of the differential thermal strains of the materials within the CIGS solar panels. Subsequently, a recovery process involves annealing the CIGS layers to remove Se, followed by leaching with nitric acid and individual extraction of the valuable metals. The study thoroughly investigates the pH values, extractant and stripping agent concentrations, organic-aqueous ratios, and reaction time to optimize the separation conditions for Cu, In, and Ga. The extraction process starts by selectively extracting In into the organic phase using D2EHPA, while Cu and Ga remain in the aqueous phase. Control of the extraction conditions enables the separation of In from Cu and Ga. Once In is extracted, Ga is subsequently extracted using the same extraction agent under different conditions, leaving nearly pure Cu in the residual aqueous solution. Ammonium hydroxide is then added to the three solutions to precipitate metal hydroxides. Under optimal conditions, a recovery rate of over 90% can be achieved for In, Ga, and Cu. Furthermore, all the formed metal hydroxides can be recycled and converted into metal oxides with a purity exceeding 99%, through calcination. These findings offer a viable pathway for effectively recycling and recovering Cu, In, and Ga from waste CIGS thin-film solar panels.

It's important to note that the recycling of CIGS solar cells is still a developing field, and commercial-scale recycling technologies may vary. The specific techniques used can depend on factors such as the desired materials to be recovered, the scale of recycling operations, and economic viability.

13.2.3 Recycling Techniques of CdTe Solar Cells (Second Generation)

CdTe solar cells have become increasingly popular as a thin-film solar cell technology due to their affordability and impressive efficiency. Recycling these solar cells is crucial for managing their life cycle, reducing environmental impact, and reclaiming valuable materials. Despite containing hazardous substances such as cadmium, appropriate recycling methods can help minimize potential risks associated with CdTe solar cells. The recycling techniques of CdTe Solar Cells are similar to the processes of CIGS solar cells. Therefore, some studies are directly organized and revealed below.

Zhang X et al. introduced a novel approach for separating and recovering rare metals from cadmium telluride waste using sulfur smelting and vacuum distillation [48]. The process involves smelting cadmium telluride with sulfur to produce tellurium with high vapor pressure. Subsequently, the tellurium is separated and purified through multi-stage vacuum distillation. The findings demonstrate that the purity of the obtained tellurium can reach an impressive 99.97%, with a recovery rate exceeding 99%. The tellurium contains minimal cadmium and sulphur, with concentrations of

0.0061% and 0.0194%, respectively. Additionally, the cadmium sulphide generated during smelting can be enriched via vacuum distillation, resulting in a purity of 99.6%.

Fthenakis V M et al. focused on several key objectives: a) Cleaning the glass to remove metals and enable glass recycling; b) Separating tellurium (Te) from cadmium (Cd) and other metals and recovering Te for its value; c) Recovering Cd for reuse or effective sequestration [49]. To achieve these goals, the solar modules were crushed into fragments and then subjected to hydrometallurgical processing, which involved leaching, ion exchange separation, precipitation, and electrowinning. It was found that low-concentration sulfuric acid solutions effectively leached cadmium and tellurium from the glass matrix. Copper and tellurium were successfully separated from the cadmium solution by utilizing a series of ion exchange columns. Subsequently, Cd and Te were separately using electrowinning and reactive precipitation methods, respectively. The electrowinning process yielded cadmium sheets with a purity of 99.5% by weight, in sizes of 2 by 4 cm and 11 by 11 cm, at a remarkably low estimated cost of only 0.1 cents per watt-peak (Wp). Furthermore, the precipitation method allowed to obtain metallic Te through various reduction agents.

13.2.4 Recycling Techniques of Perovskite Solar Cells (Third Generation)

Perovskite solar cells are a type of photovoltaic technology that have garnered significant attention in solar energy research. They are named after their crystal structure, which resembles the mineral perovskite, with the general formula ABX_3. Perovskite solar cells are known for their high-power conversion efficiency and potential for low-cost production. The active layer of these solar cells is typically composed of a perovskite material, which is a hybrid organic-inorganic compound. The most-used perovskite material is methylammonium lead iodide ($CH_3NH_3PbI_3$), and several variations and compositions are being explored to improve performance and stability. In addition to Pb and I, Cs, K, and Br are also typically employed for perovskite solar cells.

Although perovskite solar cells have yet to reach commercial scale, they have gained significant attention recently due to their high efficiency and low-cost potential. However, like any electronic device, perovskite solar cells can eventually reach the end of their useful life and require recycling. Here are some steps and a literature review of recycling perovskite solar cells [50–52].

1. Pretreatment: This typically involves disassembling the solar cells and separating the components, such as the perovskite layer, electrodes, and substrates. Various techniques can be used for material recovery, including mechanical separation, chemical dissolution, and thermal treatments.
2. Leaching: Perovskite materials can be dissolved in suitable solvents to extract them from the device. This process allows for the recovery of the perovskite layer, which can be reused to produce new solar cells. Extraction methods involve using solvents like dimethylformamide (DMF) or dimethyl sulfoxide (DMSO) to dissolve the perovskite layer, leaving behind the other components for further recycling.

3. Metal Recovery: Perovskite solar cells also contain various metals, such as tin, lead, and gold, which can be valuable and environmentally significant. After material recovery, the separated metals can be further processed to extract and recycle them. Metal recovery techniques, including hydrometallurgical processes and smelting, can be used to obtain pure metals that can be reused in various applications.
4. Substrate Recycling: The substrates used in perovskite solar cells, such as glass or flexible plastic films, can usually be recycled separately. Glass substrates can undergo typical glass recycling processes, while plastic films can be processed through plastic recycling. Recycling the substrates allows valuable resources to be conserved, and the environmental impact of solar cell waste can be reduced.

Chen B et al. reported an end-of-life material management approach for perovskite solar modules aimed at recycling toxic lead and valuable transparent conductors to safeguard the environment and yield significant economic benefits through material reuse [51]. The process involves the separation of lead from decommissioned modules using weakly acidic cation exchange resin. The separated lead can be released as soluble $Pb(NO3)_2$ and then precipitated as PbI_2 for reuse, achieving a recycling efficiency of 99.2%. Additionally, thermal delamination is employed to disassemble the encapsulated modules while preserving the integrity of transparent conductors and cover glasses. The recycled lead iodide and transparent conductors are then utilized in refabricated devices that perform similarly to those based on fresh raw materials. Moreover, a cost analysis demonstrates the economic attractiveness of this recycling technology.

In addition to recovering materials from waste perovskite solar cells, Kim B J et al. explored a novel method to manage them [53]. A straightforward and efficient technique has been developed to remove the perovskite layer and recover the mesoporous TiO_2-coated transparent conducting glass substrate for reuse. By selectively dissolving the perovskite layer, it was observed that it readily decomposes in polar aprotic solvents, due to the reaction between the solvents and Pb^{2+} cations. Remarkably, even after undergoing 10 recycling cycles, a perovskite solar cell based on a mesoporous TiO_2-coated transparent conducting glass substrate maintained a consistent power-conversion efficiency, indicating the promising potential for perovskite solar cell recycling.

It should be noticed that the materials of perovskite solar cells are still improved. Therefore, the recycling techniques for perovskite solar cells need to optimize as the technology continues to evolve. Research and development efforts are ongoing to enhance the recyclability and sustainability of perovskite solar cells to minimize their environmental impact.

REFERENCES

1. Al-Ezzi A S and Ansari M N M 2022 Photovoltaic solar cells: A review *Appl. Syst. Innov.* **5** 67.
2. Bhatia S C 2014 *Introduction to Advanced Renewable Energy Systems.* Woodhead Publishing Ltd.

3. Soga T 2006 *Nanostructured Materials for Solar Energy Conversion*. Elsevier Science Limited.
4. Breitenstein O 2013 The physics of industrial crystalline silicon solar cells. *Semiconductors and Semimetals*. **89** 1–75.
5. Tian H and Sun L 2013 Organic photovoltaics and dye-sensitized solar cells. 567–605.
6. Pastuszak J and Węgierek P 2022 Photovoltaic cell generations and current research directions for their development *Materials* **15** 554.
7. Tsakalakos L 2010 *Nanotechnology for Photovoltaics*. CRC Press.
8. Almosni S, Delamarre A, Jehl Z, Suchet D, Cojocaru L, Giteau M, et al. 2018 Material challenges for solar cells in the twenty-first century: directions in emerging technologies *Sci. Technol. Adv. Mater.* **19** 336–6.
9. Dambhare M V, Butey B and Moharil S V 2021 Solar photovoltaic technology: A review of different types of solar cells and its future trends *J. Phys. Conf. Ser.* **1913** 012053.
10. Luque A 2010 *Handbook of Photovoltaic Science and Engineering*. John Wiley & Sons.
11. Green M A 2002 Third generation photovoltaics: Solar cells for 2020 and beyond *Phys. E: Low-Dimens. Syst. Nanostruct.* **14** 65–70.
12. Ragoussi M-E and Torres T 2015 ChemInform abstract: New generation solar cells: Concepts, trends and perspectives *Chem. Commun.* **51**(19), 3957–3972.
13. Chen P-H, Chen W-S, Lee C-H and Wu J-Y 2023 Comprehensive review of crystalline silicon solar panel recycling: From historical context to advanced techniques *Sustainability* **16**(1), 60.
14. Chopra K L, Paulson P D and Dutta V 2004 Thin-film solar cells: An overview *Prog. Photovolt.* **12** 69–92.
15. Poortmans J 2006 *Thin Film Solar Cells*. John Wiley & Sons
16. Musazade E, Voloshin R, Brady N, Mondal J, Atashova S, Zharmukhamedov S K, et al. 2018 Biohybrid solar cells: Fundamentals, progress, and challenges *J. Photochem. Photobiol. C* **35** 134–56.
17. Jadoun S and Riaz U 2021 *Biohybrid solar cells Fundamentals of Solar Cell Design*. John Wiley & Sons, Inc, 117–36.
18. Nazeeruddin M D K, Baranoff E and Grätzel M 2011 Dye-sensitized solar cells: A brief overview *Sol. Energy* **85** 1172–8.
19. Lau K K S and Soroush M 2019 Overview of dye-sensitized solar cells Dye-Sensitized Solar Cells. Elsevier Inc. 1–49
20. Almora O, Roca Ldice V and Belmonte G G 2017 Perovskite solar cells: A brief introduction and some remarks. *Revista cubana de fisica* **34**(1), 58–68.
21. Thankappan A 2018 *Perovskite Photovoltaics*. Academic Press.
22. Rafiee M, Chandra S, Ahmed H and McCormack S J 2019 An overview of various configurations of luminescent solar concentrators for photovoltaic applications *Opt. Mater.* **91** 212–27.
23. Papakonstantinou I, Portnoi M and Debije M G 2021 The hidden potential of luminescent solar concentrators *Adv. Energy Mater.* **11** 2002883.
24. Nazir S, Ali A, Aftab A, Muqeet H A, Mirsaeidi S, Zhang J-M 2023 Techno-economic and environmental perspectives of solar cell technologies: A comprehensive review *Energies* **16**(13), 4959.
25. Heath G A, Silverman T J, Kempe M, Deceglie M, Ravikumar D, Remo T, et al. 2020 Research and development priorities for silicon photovoltaic module recycling to support a circular economy *Nat. Energy* **5** 502–10.
26. Walzberg J, Carpenter A and Heath G A 2021 Role of the social factors in success of solar photovoltaic reuse and recycle programmes *Nat. Energy* **6** 913–24.

27. Tasnim S S, Rahman M D M, Hasan M M, Shammi M and Tareq S M Current challenges and future perspectives of solar-PV cell waste in Bangladesh. *Heliyon* **8** e0897.
28. Azeumo M F, Germana C, Ippolito N M, Franco M, Luigi P and Settimio S Photovoltaic module recycling, a physical and a chemical recovery process *Sol. Energy Mater. Sol. Cells* **193** 314–9.
29. Maani T, Celik I, Heben M J, Ellingson R J and Apul D 2020 Environmental impacts of recycling crystalline silicon (c-SI) and cadmium telluride (CDTE) solar panels *Sci. Total Environ.* **735** 138827.
30. Cui H, Heath G, Remo T, Ravikumar D, Silverman T, Deceglie M, et al. 2022 Technoeconomic analysis of high-value, crystalline silicon photovoltaic module recycling processes *Sol. Energy Mater. Sol. Cells* **238** 111592.
31. Sah D, Chitra and Kumar S 2022 Recovery and analysis of valuable materials from a discarded crystalline silicon solar module *Sol. Energy Mater. Sol. Cells* **246** 111908.
32. Divya A, Adish T, Kaustubh P and Zade P S 2023 Review on recycling of solar modules/panels *Sol. Energy Mater. Sol. Cells* **253** 112151.
33. Tembo P M and Subramanian V 2023 Current trends in silicon-based photovoltaic recycling: A technology, assessment, and policy review *Sol. Energy* **259** 137–5.
34. Klugmann-Radziemska E, Ostrowski P, Drabczyk K, Panek P and Szkodo M 2010 Experimental validation of crystalline silicon solar cells recycling by thermal and chemical methods *Sol. Energy Mater. Sol. Cells* **94** 2275–82.
35. Kang S, Yoo S, Lee J, Boo B and Ryu H 2012 Experimental investigations for recycling of silicon and glass from waste photovoltaic modules *Renew. Energy* **47** 152–9.
36. Lee C-H, Chang Y-W, Popuri S R, Hung C-E, Liao C-H, Chang J-E, et al. 2018 Recovery of silicon, copper and aluminum from scrap silicon wafers by leaching and precipitation *Environ. Eng. Manag. J.* **17** 561–8.
37. Jung B, Park J, Seo D and Park N 2016 Sustainable system for raw-metal recovery from crystalline silicon solar panels: From noble-metal extraction to lead removal *ACS Sustain. Chem. Eng.* **4** 4079–83.
38. Shin J, Park J and Park N 2017 A method to recycle silicon wafer from end-of-life photovoltaic module and solar panels by using recycled silicon wafers *Sol. Energy Mater. Sol. Cells* **162** 1–6.
39. Yousef S, Tatariants M, Denafas J, Makarevicius V, Lukošiūtė S-I and Kruopienė J 2019 Sustainable industrial technology for recovery of Al nanocrystals, Si microparticles and Ag from solar cell wafer production waste *Sol. Energy Mater. Sol. Cells* **191** 493–50.
40. Lee J-S, Ahn Y-S, Kang G-H and Wang J-P 2016 Recovery of 4N-grade copper from photovoltaic ribbon in spent solar module *Mater. Technol.* **31** 574.
41. Moon G and Yoo K 2017 Separation of Cu, Sn, Pb from photovoltaic ribbon by hydrochloric acid leaching with stannic ion followed by solvent extraction *Hydrometallurgy* **171** 123–7. Available from: 10.1016/j.hydromet.2017.05.003
42. Chen W-S, Chen Y-J, Lee C-H, Cheng Y-J, Chen Y-A, Liu F-W, et al. 2021 Recovery of valuable materials from the waste crystalline-silicon photovoltaic cell and ribbon *Processes* **9** 712.
43. Teknetzi I, Holgersson S and Ebin B 2023 Valuable metal recycling from thin film CIGS solar cells by leaching under mild conditions *Sol. Energy Mater. Sol. Cells* **252** 112178.
44. M K Gustafsson, A Mougas 2014 Recycling of CIGS solar cell waste materials. Chalmers Tekniska Hogskola (Sweden).
45. Gustafsson A M K, Foreman M R StJ and Ekberg C 2014 Recycling of high purity selenium from CIGS solar cell waste materials *Waste Manag.* **34** 1775–82.

46. Zimmermann Y-S, Niewersch C, Lenz M, Kül Z Z, Corvini P F-X, Schäffer A, et al. 2014 Recycling of indium from CIGS photovoltaic cells: potential of combining acid-resistant nanofiltration with liquid–liquid extraction *Environ. Sci. Technol.* **48** 13412–8.
47. Liu F-W, Cheng T-M, Chen Y-J, Yueh K-C, Tang S-Y, Wang K, et al. 2022 High-yield recycling and recovery of copper, indium, and gallium from waste copper indium gallium selenide thin-film solar panels *Sol. Energy Mater. Sol. Cells* **241** 111691.
48. Zhang X, Liu D, Jiang W, Xu W, Deng P, Deng J, et al. 2020 Application of multi-stage vacuum distillation for secondary resource recovery: potential recovery method of cadmium telluride photovoltaic waste *J. Mater. Res. Technol.* **9** 6977–86.
49. Fthenakis V M, Duby P F, Wang W, Graves C R and Belova A A 2006 *Recycling of CdTe Photovoltaic Modules: Recovery of Cadmium and Tellurium.* 21st European photovoltaic solar energy conference
50. Chhillar P, Dhamaniya B P, Dutta V and Pathak S K 2019 Recycling of perovskite films: Route toward cost-efficient and environment-friendly perovskite technology *ACS Omega* **4** 11880–7.
51. Chen B, Fei C, Chen S, Gu H, Xiao X and Huang J 2021 Recycling lead and transparent conductors from perovskite solar modules *Nat. Commun.* **12**(1), 5859–5868.
52. Akulenko E S, Hadadian M, Santasalo-Aarnio A and Miettunen K 2023 Eco-design for perovskite solar cells to address future waste challenges and recover valuable materials *Heliyon* **9** e13584.
53. Kim B J, Kim D H, Kwon S L, Park S Y, Li Z, Zhu K, et al. 2016 Selective dissolution of halide perovskites as a step towards recycling solar cells *Nat. Commun.* **7**(1), 1–9.

3. Direct Methanol Fuel Cell: DMFC is a fuel cell that operates at relatively low temperatures and offers high energy density. DMFCs are known for their ability to use liquid methanol as a fuel, which eliminates the need for hydrogen gas storage and allows for easy refueling. Within the DMFC, methanol is supplied to the anode, where it undergoes oxidation in the presence of a catalyst, typically platinum. The oxidation reaction produces carbon dioxide, protons (H^+), and electrons (e^-). The protons migrate through a proton exchange membrane, while the electrons are forced to travel through an external circuit, generating electric current. At the cathode, oxygen from the air reacts with the protons and electrons to produce water as the main byproduct. DMFCs have a compact design, making them suitable for portable applications, such as small electronic devices, portable power banks, and backup power systems. They offer advantages like high energy density, quick startup times, and low emissions. However, challenges such as methanol crossover and the cost of catalyst materials are areas of ongoing research and development to improve DMFC technology [20, 21].
4. Alkaline Fuel Cell: AFC is a fuel cell that uses an alkaline electrolyte, typically potassium hydroxide (KOH), to facilitate the electrochemical reaction and generate electrical energy. AFCs were one of the earliest types of fuel cells developed and have a long history of use in various applications. In an AFC, hydrogen gas is supplied to the anode (negative electrode), while oxygen or air is provided to the cathode (positive electrode). The alkaline electrolyte allows hydroxide ions (OH^-) to flow between the electrodes, while a porous electrode structure and a catalyst, often made of platinum or other metals, facilitate the electrochemical reactions. As hydrogen at the anode combines with hydroxide ions, water and electrons are produced. The electrons flow through an external circuit, generating an electric current, while the hydroxide ions migrate through the electrolyte to the cathode. At the cathode, oxygen and hydroxide ions combine to produce water as the main byproduct. AFCs have been used in various applications, including space missions, due to their high efficiency and reliability. However, they are sensitive to carbon dioxide and require pure hydrogen and oxygen, making fuel purification necessary. Ongoing research aims to enhance the performance and durability of AFCs and explore their potential in emerging energy systems [22, 23].
5. Phosphoric Acid Fuel Cell: PAFC is a fuel cell that operates through an electrochemical reaction using phosphoric acid as the electrolyte. PAFCs are known for their durability, high efficiency, and ability to operate at moderate temperatures. In a PAFC, hydrogen gas is supplied to the anode (negative electrode), where it is oxidized in the presence of a catalyst, typically platinum or a platinum-ruthenium alloy. This oxidation reaction generates protons (H^+) and electrons (e^-). The protons pass through the phosphoric acid electrolyte, while the electrons flow through an external circuit, producing electric current. At the cathode, oxygen from the air combines with the protons and electrons, forming water as the main byproduct.

The phosphoric acid electrolyte helps to transfer protons and maintain the reaction. PAFCs operate at temperatures typically ranging from 150 to 200 degrees Celsius, making them suitable for stationary power generation applications. They have been used in buildings, hospitals, and other large-scale power generation systems. The use of phosphoric acid as the electrolyte offers advantages such as durability and resistance to fuel impurities. However, PAFCs require a fuel reformer to convert hydrocarbon fuels into hydrogen, and their operating temperature limits their ability to start up quickly. Research efforts are focused on improving their performance, reducing costs, and exploring their potential in combined heat and power applications [24, 25].

6. Molten Carbonate Fuel Cell: MCFCs are known for their ability to generate electricity with high efficiency and can utilize a variety of fuels, including natural gas, biogas, and even coal-derived gases. In an MCFC, fuel, typically a hydrocarbon-based gas, is supplied to the anode, while oxygen or air is provided to the cathode. The electrolyte used in MCFCs is a mixture of lithium and potassium carbonates, which is in a molten state at high temperatures, typically around 650 to 750 degrees Celsius. The elevated operating temperature allows for efficient ion conduction and reaction kinetics. At the anode, the fuel undergoes oxidation, releasing carbon dioxide, electrons, and carbonate ions (CO_3^{2-}). The electrons flow through an external circuit, producing an electric current, while the carbonate ions migrate through the molten electrolyte to the cathode. At the cathode, oxygen combines with the carbonate ions and electrons to form carbonate ions again. MCFCs are primarily used in large-scale power generation applications, such as industrial and utility power plants. They offer advantages like high efficiency, fuel flexibility, and the potential for combined heat and power generation. The high operating temperature also allows MCFCs to internally reform hydrocarbon fuels, simplifying the fuel processing requirements. However, the high operating temperature poses challenges regarding materials selection, system design, and durability [26, 27].
7. Reversible/Regenerative Fuel Cell: RFC is a unique type of fuel cell that can function in both fuel cell and electrolysis modes, enabling it to convert electrical energy into chemical energy and vice versa. The RFC functions similarly to other fuel cells in the fuel cell mode. It utilizes a fuel, such as hydrogen, at the anode and an oxidizer, typically oxygen or air, at the cathode. Through electrochemical reactions, the fuel is oxidized, generating electricity, water, and heat as byproducts. The generated electricity can be utilized to power various applications. In the electrolysis mode, the RFC reverses its operation, using electrical energy to split water into hydrogen and oxygen, allowing for energy storage in the form of hydrogen. RFCs have the potential to play a crucial role in energy storage and renewable energy systems by facilitating the efficient conversion and storage of electricity as hydrogen, which can later be used as a fuel source for electricity generation [28, 29]. The brief comparison of different fuel cells is revealed in Table 14.1 [30].

TABLE 14.1
Comparison of Different Fuel Cells [30]

	Operating Temperature (°C)	Power (kW)	Efficiency (%)
PEMFC	60–110	0.01–250	40–55
SOFC	500–1000	0.5–2000	40–72
DMFC	70–130	0.001–100	40
AFC	70–130	0.1–50	50–70
PAFC	150–200	50–1000	40–45
MCFC	650–750	200–100,000	50–60

14.2 RECYCLING TECHNIQUES OF FUEL CELLS

Fuel cells are in the early stages of research and development, and their widespread adoption and commercialization has not happened. Therefore, there is little information on the current condition and quantity of waste fuel cells. Nevertheless, to prevent wastes from producing in the future, many scholars have already developed recycling techniques for different types of fuel cells (The accumulation and recovery of precious ingredients, including precious metals, from a significant number of fuel cells reaching the end of their useful life in 10–15 years, are of utmost importance [31]). Recycling fuel cells is essential to promote sustainability, conserve valuable resources, and reduce environmental impacts. Various recycling techniques are employed to recover materials from used fuel cells. The specific recycling process can depend on the type of fuel cell technology and the materials used in its construction. Below are some common steps to recycle materials from fuel cells. The recycling techniques of typical fuel cells will also be introduced [32].

1. Material Separation: The first step in recycling fuel cells involves dismantling and disassembling the units. Different components, such as electrodes, electrolytes, and casings, need to be separated to facilitate the recovery of valuable materials. Also, shredding, milling, and sieving procedures can also roughly separate metals from other materials such as plastic.
2. Metal Recycling: This step mainly involves pyrometallurgical, hydrometallurgical, and electrochemical processes. In the pyrometallurgical procedure, the fuel cell components are subjected to high-temperature processing (typically in a furnace) to separate metals and other materials. This technique can be used for certain types of fuel cells, like SOFCs that use metal-based materials. The hydrometallurgical technique involves using chemical processes, such as leaching and precipitation, to recover metals and materials from fuel cell components. This method is often used for fuel cells containing valuable metals like platinum, which is commonly used as a catalyst in PEMFCs. On the other hand, some fuel cell recycling processes employ electrochemical methods to dissolve and recover metals from various components. Electrolysis and other electrochemical processes can be used for selective metal recovery.

3. Refurbishment and Reuse: In some cases, used fuel cell components may undergo refurbishment to extend their lifespan or be repurposed in other applications.

14.2.1 Recycling Techniques of PEMFC

Recycling techniques for Proton Exchange Membrane fuel cells (PEMFC) include:

1. Dismantling and Disassembly: The first step in recycling PEMFCs is dismantling and disassembling the fuel cell stack. This process involves removing the catalyst layers, membranes, gas diffusion layers, and other components. Care must be taken during this step to avoid damage to the individual parts.
2. Electrolyte Membrane Recycling: The proton exchange membrane (PEM) is a crucial component of the fuel cell. Recycling techniques for the PEM involve several methods, such as solvent extraction, pyrolysis, and mechanical processes. Solvent extraction can recover the polymer for reuse, while pyrolysis breaks down the membrane into smaller molecules that can be used as fuel or raw materials for other processes. Besides, the membrane electrode assembly (MEA) of a PEM fuel cell can be recycled through a solvent-based approach involving a lower alkyl alcohol solvent. This method allows separation of MEA components without resorting to combustion or removing outer layers, such as gas diffusion layers.
3. Catalyst Recovery: The catalyst in a PEMFC typically consists of precious metals like platinum, which are expensive and valuable. Catalyst recovery methods involve different chemical and electrochemical processes to extract and purify the precious metals for reuse in other fuel cells or industrial applications.
4. Gas Diffusion Layer Recovery: The gas diffusion layer (GDL) is typically made of carbon-based materials. The GDL can be recovered by various methods, including thermal treatment to remove contaminants and impurities, and then grinding or milling the material to produce carbon particles or carbon black, which can be used in other applications.
5. Bipolar Plate Recycling: The bipolar plates in a PEMFC are usually made of conductive graphite or metal. The recycling process for bipolar plates involves removing any coatings or contaminants and reprocessing the material for use in other applications.
6. Reconditioning and Refurbishing: In some cases, rather than complete recycling, PEMFC components can be reconditioned and refurbished for extended use. This involves testing and repairing any damaged parts to bring the fuel cell back to working condition.

Zhao J et al. designed several steps to reclaim platinum from Pt/C catalysts in PEMFC [33]. First, the degraded Pt/C catalysts are dried at 80 °C for 3 hours and then sintered at 600 °C for 6 hours. After that, they are dissolved using aqua fortis, purified with hydrochloric acid, and reduced and filtered. The reclaimed platinum is then used to prepare Pt/C catalysts again, employing two proposed processes: pH value control and mass control. The fuel cell utilizing these recycled catalysts demonstrates a power density exceeding 0.18 W cm^{-2}. This approach can lead to a reduction in the overall cost of fuel cell technology.

Patel A at al. focused on recovering platinum from secondary sources, specifically end-of-life PEMFCs, using electrowinning and chemical dissolution in deoxygenated 4 M potassium iodide, with varying amounts of added iodine [34]. The research evaluated the leach rate of platinum black deposited on an electrochemical quartz crystal microbalance (EQCM) and the effective recovery of Pt from both untested and end-of-life PEMFCs. The dissolution rate of platinum black was observed to be influenced by the concentration of added iodine, with higher amounts accelerating the reaction. The platinum recovery from leached materials was remarkably high, reaching 98.7% for untested PEMFCs and 96.7% for end-of-life PEMFCs, as determined by aqua regia digestion. The results showed that higher iodine concentrations led to continuously improved recovery efficiency, but the improvements became relatively minor beyond 5 mM of iodine.

Duclos L et al. proposed the detailed development of an efficient hydrometallurgical recovery process for platinum from Pt catalysts used in PEMFC electrodes, conducted at the laboratory scale [35]. The study optimized the recovery process by combining leaching, Pt separation, precipitation, and filtration steps. Two different leaching agents (H_2O_2/HCl and HNO_3/HCl) and two separation methods (liquid-liquid extraction and ion exchange resins) were compared for their efficiencies. The recovered platinum was precipitated as ammonium hexachloroplatinate ((NH_4)2PtCl_6), a precursor for PEMFC catalysts, at the end of each process alternative. This led to the evaluation of four different process alternatives. Initially, experiments were performed using Pt/C particles and hexachloroplatinic acid to study the impact of various experimental parameters on platinum recovery efficiency. Subsequently, the optimized leaching and recovery process was applied to real PEMFC catalyst coated membranes (CCMs). The most efficient process alternative resulted in a 76% platinum recovery yield from the CCMs.

Duclos L et al. developed an eco-friendly recycling method to selectively recover Co and Pt from a Pt_3Co/C catalyst utilized in PEMFC cathodes [36]. Two distinct hydrometallurgical approaches were employed for Pt and Co separation, involving ion exchange resin (Lewatit-MP62 resin) and solvent extraction (Cyanex 923). The recycling process was fine-tuned to maximize platinum recovery, resulting in yields of 85% and 78% for Pt from the membrane-electrode assembly (MEA) using solvent extraction and ion exchange resin separation, respectively. The recovered platinum alkaline solutions were then utilized for synthesizing Pt/C particles via a modified polyol method. Although the resin alternative yielded slightly less recovery, the prepared catalyst displayed satisfactory morphological and electrochemical properties for the oxygen reduction reaction (ORR), performing comparably to a commercial state-of-the-art Pt/C catalyst. Thus, the feasibility of closed-loop Pt/C catalyst recycling was successfully demonstrated. The environmental impacts of both recycling alternatives were assessed through life cycle assessment (LCA), and they were found to be similar from the perspective of the entire MEA life cycle. Consequently, the most suitable option was the ion exchange resin alternative.

Chen W-S et al. mentioned that PEMFCs have emerged as a promising alternative, but their reliance on platinum catalysts presents challenges due to limited resources and scarcity [37]. Therefore, to address this, the study proposed a highly efficient method for platinum recovery while preserving spent membranes. The separation of the membrane and catalyst was achieved using isopropanol, and the spent membrane was dissolved in a 50% ethanol solution to prepare a precursor for subsequent membrane regeneration. For platinum leaching, hydrochloric acid (HCl) was utilized as the leaching agent, and various experimental parameters like HCl concentration, H_2O_2 concentration, contact time, and operating temperature were fine-tuned to achieve the highest platinum leaching rate. Through isothermal leaching experiments, the leaching mechanism was explored using the shrinking core model, revealing the involvement of both surface chemical and inner diffusion mechanisms in the platinum leaching process, with the inner diffusion mechanism playing a primary role. The optimized conditions resulted in an impressive platinum leaching rate of approximately 90%, and the activation energy of the reaction was calculated as 6.89 kJ/mol, using the Arrhenius equation.

In addition to recovering metals from PEMFC, there are still other techniques to reach the goal of resource circulation. Zeng L et al. introduced a novel double-recycling engineering approach for regenerating Pt/C graphitized hollow nanosphere catalysts [38]. The process involved recycling Pt from waste PEMFC as a platinum source and utilizing waste pomelo peel as a carbon source. By employing polyol reduction, Pt/C catalysts were successfully regenerated, contributing to sustainable energy development. The platinum recovery efficiency reached an impressive 96.4 ± 0.2%. Graphitized hollow nanospheres were synthesized by carefully adjusting critical factors, such as pyrolysis temperature and treatment method. The waste pomelo peel material was transformed into graphitized hollow nanospheres using an improved method at 1000 °C (referred to as PP-2–1000). The recovered Pt was then combined with PP-2–1000 and commercial carbon (CC) to create Pt/PP-2–1000 and Pt/CC catalysts, respectively. Remarkably, Pt/PP-2–1000 exhibited an electrochemical active surface area (ECSA) equivalent to Pt/CC. Furthermore, for the oxygen-reduction reaction (ORR), Pt/PP-2–1000 demonstrated a mass activity (i_m at 0.9 V) that was 2.98 times higher than that of Pt/CC. Additionally, after undergoing the accelerated durability test (ADT), Pt/PP-2–1000 showed a 10.1% higher efficiency of remaining ECSA compared to Pt/CC. This significant improvement signifies the excellent stability and electrochemical performance of the recovered Pt/PP-2–1000 catalyst.

Xu F et al. conducted a test to develop a simple, highly efficient, and environmentally friendly method for recycling the essential materials of a membrane electrode assembly (MEA) used in PEMFC [39]. The process involved immersing the catalyst-coated membranes (CCMs) into sulfuric acid until a transparent solution containing Pt and perfluorosulfonic acid resin was formed. This solution caused the membrane to dissolve while also oxidizing the amorphous carbon nanoparticles serving as catalyst supports in the catalyst layers. Following this, a centrifugal separation was employed to separate the metal Pt and perfluorosulfonic acid resin. The separated resin was then reconstituted into a new membrane, and its performance was evaluated in a single fuel cell test. The results demonstrated the potential of this solution for effectively recycling the critical materials of MEAs in PEMFC, offering a promising approach for sustainable material reuse in fuel cell technology.

Shore L designed a patent that demonstrated a novel approach for recovering ionomers and noble metals from PEM devices was introduced that employed an alcohol solvent process [40]. This patent outlines a pre-treatment method that enables the subsequent retrieval of the polymer resin and the catalyst. The process initiates with the delamination of MEA layers using an alkyl alcohol solution, while a microwave heater is utilized to raise the temperature to a suitable range for dissolution. The particle size of the polymer within the dispersion plays a crucial role in the subsequent separation process, accomplished through filter-pressing and ultra-filtration. The microwave heater effectively controls this parameter by adjusting residence time and temperature, optimizing the recovery of materials from the PEM devices.

14.2.2 Recycling Techniques of SOFC

Several recycling techniques for SOFCs have been explored now. First is material recovery. Since SOFCs contain valuable materials like iron, aluminium, chromium, nickel, and other rare metals (platinum and yttrium) that can be recovered through recycling, techniques such as hydrometallurgical processes and pyrometallurgical processes are used to extract and recycle these materials from spent SOFCs. The other method is refurbishment and reuse. Some components of the SOFC stack, such as the interconnectors and support structures, can be refurbished and reused in new SOFCs. This approach helps to reduce the consumption of new materials and lowers the overall cost of manufacturing SOFCs. The detailed introduction and some cases are shown below [41].

1. Component separation: The first step in recycling SOFCs is to separate various components, such as the anode, cathode, electrolyte, and interconnect. This can be achieved through mechanical processes, heat treatment, or chemical methods.
2. Pyrolysis: Pyrolysis is a thermal process that involves heating the SOFC components in an oxygen-limited environment at high temperatures. This breaks down the organic materials into gases and leaves behind inorganic materials, such as the ceramic electrolyte and interconnect.
3. Electrolyte reclamation: Recovering the ceramic electrolyte, often made of yttria-stabilized zirconia (YSZ), is crucial for recycling. Techniques like crushing and sieving can be used to separate the YSZ from other components.
4. Metal recovery: The interconnect material in SOFCs is typically made of metals like stainless steel. Separating and recovering these metals through methods like mechanical separation or smelting is essential for reuse or recycling. For example, chemical methods, such as acid leaching or solvent extraction, can selectively dissolve and separate specific components from SOFCs. Moreover, the entire SOFC stack may also be subjected to high-temperature processes, like melting or sintering, to recover and recycle valuable materials.
5. Anode and cathode materials recovery: The anode and cathode in SOFCs may contain valuable materials like nickel and lanthanum strontium manganite (LSM). Extracting and recovering these materials can be accomplished through processes like leaching, precipitation, and chemical reduction.

Saffirio S et al. proposed research aims to address the lack of validated and scalable recycling methods for SOC technologies, specifically focusing on the recovery of Yttria-stabilized Zirconia (YSZ) from End-of-Life (EoL) SOC components [42]. To achieve this, a comprehensive recycling process has been developed, involving several steps. Initially, the EoL composite materials undergo a pre-milling process for 6 hours, followed by hydrothermal treatment (HT) at 200 °C for 4 hours to enhance the disaggregation of the sintered composite. Subsequently, selective oxidative leaching using HNO_3 solution at 80 °C for 2 hours is employed to extract Ni^{2+} from the composite Ni-YSZ materials. This optimized procedure yields YSZ powders with well-defined specifications in terms of particle size distribution, specific surface area, and chemical purity. The intermediate HT step is particularly crucial, as it promotes the disaggregation of the sintered powders, resulting in an increased specific surface area (up to 13 $m^2\ g^{-1}$) and a reduction in primary particle aggregates. The acid-assisted leaching effectively extracts Nickel from the composite Ni-YSZ powders while preserving YSZ crystallinity and showing negligible loss of Zr and Y, as confirmed by ICP analysis on the recovered supernatants.

Yenesew GT et al. presented a novel strategy for recycling and recovering components from SOFCs [43]. The process involves the separation of electrodes and electrolytes from commercial cells through mechanical scraping and grinding, followed by thermal and chemical treatments. Notably, the air electrode ($La_xSr_{1-x}CoO_3$), which accounts for a significant portion of the cell weight, along with nickel oxide (NiO) and yttria-stabilized zirconia (YSZ) from both the fuel electrode and electrolyte, were successfully separated. Approximately 90–92% of the initial Ni and YSZ weight can be collected during this recycling process. The recovered materials undergo thorough characterization through X-ray diffraction, scanning electron microscopy, (thermal, chemical) analysis, and BET surface area measurement, with impurities remaining below 1 At.%. The recycled YSZ materials exhibit comparable conductivity levels to commercial YSZ, with a total value exceeding 6.4×10^{-3} S cm^{-1} at 700 °C. Moreover, the contributions of grain and grain boundary conductivities to the total conductivity are distinguishable, particularly at lower temperatures. This research demonstrates a promising approach for efficiently recycling and reusing SOFC ceramic components, thereby contributing to sustainable and environmentally friendly energy technologies.

14.2.3 Recycling Techniques of AFC

As mentioned above, the development of fuel cells is still progressing. Hence, the recycling techniques of AFC are also being explored. Below are some general techniques and methods for resource circulation from AFC.

1. Recovering precious metals: A recycling process has been designed to recover platinum from the membrane electrode assembly (MEA) of AFCs. This process focuses on extracting valuable materials from the fuel cell components.
2. Rolling technique: In the recycling of AFC components, porous carbon electrodes containing platinum and platinum-ruthenium catalysts can be prepared

using a rolling technique with polytetrafluoroethylene as a binder. This technique allows for the reuse of precious metals in the electrodes.

3. Integration with elastocaloric cooling: One study proposes a hybrid system model that integrates AFCs with temperature-matching elastocaloric coolers to recycle the byproduct waste heat for cooling production. This approach aims to utilize the waste heat generated by AFCs for other purposes [44, 45].

It's important to understand that while these recycling techniques are being developed and implemented, there is still ongoing research and innovation in fuel cell recycling. The aim is to improve the recycling efficiency and sustainability of the recycling process, reduce pollution, and promote the circular economy.

REFERENCES

1. Chellappan M V, Todorovic M H and Enjeti P N 2008 Fuel Cell Based Battery-Less UPS System. In 2008 *IEEE Industry Applications Society Annual Meeting* (pp. 1–8). IEEE.
2. Breeze P 2017 *Fuel Cells*. Academic Press.
3. Pan Z F, An L and Wen C Y 2019 Recent advances in fuel cells based propulsion systems for unmanned aerial vehicles *Appl. Energy* **240** 473–85.
4. Edwards P P, Kuznetsov V L, David W I F and Brandon N P 2018 Hydrogen and fuel cells: Towards a sustainable energy future *Energy Policy* **36** 4356–62.
5. Neef H-J 2009 International overview of hydrogen and fuel cell research *Energy* **34** 327–33.
6. Carrette L, Friedrich K A and Stimming U 2000 Fuel cells: Principles, types, fuels, and applications *ChemPhysChem* **1** 162–93.
7. Carrette L, Friedrich K A and Stimming U 2001 Fuel cells - Fundamentals and applications *Fuel Cells* **1** 5–39.
8. Garche J, Jurissen L 2015 Applications of fuel cell technology: Status and perspectives *Electrochem. Soc. Interface*. **24** 39–43.
9. Luo Y, Wu Y, Li B, Mo T, Li Y, Feng S-P, et al. Development and application of fuel cells in the automobile industry *J. Energy Storage* **42** 103124.
10. Perčić M, Vladimir N, Jovanović I and Koričan M Application of fuel cells with zero-carbon fuels in short-sea shipping *Appl. Energy* **309** 11846.
11. Preli F 2002 Technical challenges for fuel cells in mobile applications *Fuel Cells* **2** 5–9.
12. Herring A M 2010 *Fuel Cell Chemistry and Operation*. OUP USA.
13. Wang J, Wang H and Fan Y 2018 Techno-economic challenges of fuel cell commercialization *Engineering* **4** 352–60.
14. Gamburzev S and Appleby A J 2002 Recent progress in performance improvement of the proton exchange membrane fuel cell (PEMFC) *J. Power Sources* **107** 5–12.
15. Ren X, Wang Y, Liu A, Zhang Z, Lv Q and Liu B 2020 Current progress and performance improvement of Pt/C catalysts for fuel cells *J. Mater. Chem. A* **8** 24284–306.
16. Yang C, Costamagna P, Srinivasan S, Benziger J and Bocarsly A B 2001 Approaches and technical challenges to high temperature operation of proton exchange membrane fuel cells *J. Power Sources* **103** 1–9.
17. Wang C, Wang S, Peng L, Zhang J, Shao Z, Huang J, et al. 2016 Recent progress on the key materials and components for proton exchange membrane fuel cells in vehicle applications *Energies* **9** 603.
18. Minh N 2004 Solid oxide fuel cell technology?features and applications *Solid State Ion.* **174** 271–7.

19. Cooper S J and Brandon N P 2017 An introduction to solid oxide fuel cell materials, technology and applications. Solid Oxide Fuel Cell Lifetime and Reliability. Elsevier Ltd. 1–18.
20. Kamarudin S K, Daud W R W, Ho S L and Hasran U A 2007 Overview on the challenges and developments of micro-direct methanol fuel cells (DMFC) *J. Power Sources* **163** 743–54.
21. Alias M S, Kamarudin S K, Zainoodin A M and Masdar M S 2020 Active direct methanol fuel cell: An overview *Int. J. Hydrog. Energy* **45** 19620–41.
22. McLean G F, Niet T, Prince-Richard S and Djilali N 2005 An assessment of alkaline fuel cell technology. *International Journal of Hydrogen Energy* **27**(5), 507–526.
23. Wang Y, Leung D Y C, Xuan J and Wang H 2017 A review on unitized regenerative fuel cell technologies, part B: Unitized regenerative alkaline fuel cell, solid oxide fuel cell, and microfluidic fuel cell *Renew. Sust. Energ. Rev.* **75** 775–95.
24. Sammes N, Bove R and Stahl K 2004 Phosphoric acid fuel cells: Fundamentals and applications *Curr. Opin. Solid State Mater. Sci.* **8** 372–8.
25. Eapen D E, Suseendiran S R and Rengaswamy R 2016 Phosphoric acid fuel cells. *Compendium of Hydrogen Energy Volume 3: Hydrogen Energy Conversion*. Woodhead Publishing.
26. Watanabe T, Izaki Y, Mugikura Y, Morita H, Yoshikawa M, Kawase M, et al. 2006 Applicability of molten carbonate fuel cells to various fuels *J. Power Sources* **160** 868–71.
27. Mehrpooya M, Bahramian P, Pourfayaz F and Rosen M A 2016 Introducing and analysis of a hybrid molten carbonate fuel cell-supercritical carbon dioxide Brayton cycle system *Sustain. Energy Technol. Assess.* **18** 100–6.
28. O'Hayre R P 2016 *Fuel Cell Fundamentals*. John Wiley & Sons.
29. Pu Z, Zhang G, Hassanpour A, Zheng D, Wang S, Liao S, et al. 2021 Regenerative fuel cells: Recent progress, challenges, perspectives and their applications for space energy system *Appl. Energy* **283** 116376.
30. Irshad M, Siraj K, Raza R, Ali A, Tiwari P, Zhu B, et al. A brief description of high temperature solid oxide fuel cell's operation, materials, design, fabrication technologies and performance *Appl. Sci.* **6** 75.
31. Song H 2022 *Nanotechnology in Fuel Cells*. Elsevier.
32. Valente A, Iribarren D and Dufour J 2019 End of life of fuel cells and hydrogen products: From technologies to strategies *Int. J. Hydrog. Energy* **44** 20965–77.
33. Zhao J, He X, Tian J, Wan C and Jiang C Reclaim/recycle of Pt/C catalysts for PEMFC *Energy Convers. Manag.* **48** 450–3.
34. Patel A and Dawson R 2015 Recovery of platinum group metal value via potassium iodide leaching *Hydrometallurgy* **157** 219–25.
35. Duclos L, Svecova L, Laforest V, Mandil G and Thivel P-X 2016 Process development and optimization for platinum recovery from PEM fuel cell catalyst *Hydrometallurgy* **160** 79–89.
36. Duclos L, Chattot R, Dubau L, Thivel P-X, Mandil G, Laforest V, et al. 2020 Closing the loop: Life cycle assessment and optimization of a PEMFC platinum-based catalyst recycling process *Green Chem.* **22** 1919–33.
37. Chen W-S, Liu W-S and Chen W-C 2023 Leaching efficiency and kinetics of platinum from spent proton exchange membrane fuel cells by H_2O_2/HCl *Metals* **13** 1006.
38. Zeng L, Wang Y, Nhung N T H, He C, Chen H, Wang X, et al. 2022 Double recovery and regeneration of Pt/C catalysts: Both platinum from the spent proton exchange membrane fuel cell stacks and carbon from the pomelo peel *Electrochim. Acta* **428** 140918.
39. Xu F, Mu S and Pan M 2010 Recycling of membrane electrode assembly of PEMFC by acid *Int. J. Hydrog. Energy* **35** 2976.

40. Shore L 2012 U.S. Patent No. 8124,261 Washington, DC U.S. Patent and Trademark Office.
41. Sarner S, Schreiber A, Menzler N H and Guillon O 2022 Recycling strategies for solid oxide cells *Adv. Energy Mater.* **12** 2201805.
42. Saffirio S, Pylypko S, Fiorot S, Schiavi I, Fiore S, Santarelli M, et al. Hydrothermally-assisted recovery of yttria-stabilized zirconia (YSZ) from end-of-life solid oxide cells 2022 *SM&T* **33** e00473.
43. Yenesew G T, Quarez E, Le Gal La Salle A, Nicollet C and Joubert O 2023 Recycling and characterization of end-of-life solid oxide fuel/electrolyzer ceramic material cell components *Resour. Conserv. Recycl.* **190** 106809.
44. Zhang X, Cai L, Liao T, Zhou Y, Zhao Y and Chen J 2018 Exploiting the waste heat from an alkaline fuel cell via electrochemical cycles *Energy* **142** 983–90.
45. Zhu H, Li J, Lai C and Zhang H 2022 Recycling alkaline fuel cell waste heat for cooling production via temperature-matching elastocaloric cooler *Int. J. Hydrog. Energy* **47** 27124–38.

15 Concluding Remarks

Ngoc Thanh Thuy Tran, Shou-Heng Liu, Wei-Sheng Chen, Chia-Yu Lin, and Wen-Hui Cheng
National Cheng Kung University, Tainan City, Taiwan

Han-Yi Chen
National Tsing Hua University, Hsinchu City, Taiwan

Chih-Ming Kao
Institute of Environmental Engineering, National Sun Yat-sen University, Kaohsiung City, Taiwan

Shu-Yi Tsai
National Cheng Kung University, Tainan City, Taiwan

Hsing-Jung Ho
Tohoku University, Sendai City, Japan

Jow-Lay Huang
National Cheng Kung University, Tainan City, Taiwan

In this book, the collective insights gathered from the diverse chapters leave us with a profound understanding of the intricate relationship between circular economy principles and sustainable energy practices. The journey through carbon capture, mineral carbonation, biomass energy, and the recycling of waste materials has illuminated the path toward a more harmonious coexistence between economic growth and ecological preservation. By fostering a culture of resource efficiency, waste reduction, and the integration of renewable energy, this book presents a blueprint for transformative change in our global approach to energy and resource management. As we envision a future characterized by circularity, sustainability, and net-zero emissions, the chapters within this volume stand as a testament to the collaborative efforts needed to forge a resilient and environmentally conscious world for generations to come.

This book offers a compelling perspective on the dynamic interplay between circular economy principles and sustainable energy solutions. The diverse chapters collectively underscore the urgency of reimagining our approach to resource management, waste reduction, and renewable energy utilization. This book not only

DOI: 10.1201/9781003463009-15

serves as a snapshot of the current state of affairs but also as a guide for navigating the complexities of the environmental challenges we face. The perspective gained from these insights emphasizes the transformative power of circular economy strategies, encouraging a shift towards more responsible and sustainable practices across industries.

The potential for achieving net-zero emissions in the agricultural and industrial sectors through various technologies, innovations, and practices and their integration with CE strategies was discussed in Chapter 2. Although achieving net-zero emissions through more mature technologies and practices (commercial or pilot-phase solutions) is feasible, the lack of effective policies, and high costs, especially in developing countries should be improved. To overcome these challenges, it is crucial to remove adoption barriers and implement strategies that promote the widespread adoption of sustainable practices. Further research is needed to make these technologies economically viable and scalable, while understanding their socio-environmental impacts. Also, the interlinking the pathways to net-zero emissions and combining them with circular and demand-side measures are important. Further research is needed to determine the optimal combinations of net-zero pathways to minimize resource scarcity caused by decarbonizing agriculture and industry. Innovative technologies combined with new business models will provide opportunities to change the current linear production and consumption patterns, creating higher benefits and value for society through resource utilization. These ideas provide a blueprint for future research and development of CE policies and technologies, giving investors, consumers, researchers, and policy regulatory agencies a broader understanding of CE models.

Carbon capture is expected to play a crucial role in mitigating climate change and achieving greenhouse gas reduction targets. It can achieve the goals such as below.

1. Greenhouse Gas Reduction: Carbon capture technologies help reduce greenhouse gas emissions by capturing and storing CO_2 from industrial processes that would otherwise be released into the atmosphere.
2. Transition to a Low-Carbon Economy: Carbon capture can facilitate the transition to a low-carbon economy by providing a bridge technology that allows for the gradual phase-out of fossil fuels.
3. Carbon Removal and Negative Emissions: Carbon capture technologies are also being explored for carbon removal and negative emissions strategies. These approaches involve removing CO_2 from the atmosphere to actively reduce its concentration.

While carbon capture holds significant potential, there are still challenges to overcome, including scaling up the technology, ensuring long-term storage integrity, and addressing the energy requirements and environmental impacts associated with the capture process as mentioned in Chapter 3. However, as the urgency to address climate change grows, it is expected that carbon capture will continue to evolve and play a vital role in reducing greenhouse gas emissions and achieving a sustainable future.

Mineral carbonation is promising technology that can capture and utilize CO_2 as stable carbonates. Using industrial wastes as raw materials for mineral carbonation technology can not only reduce CO_2 emissions, but also avoid the negative environmental impacts from the waste. The industries that generate alkaline waste usually emit large quantities of CO_2 flue gas, such as cement and steel industries. Hence, mineral carbonation of industrial waste is regarded as a promising route that reduces unavoidable industrial emissions from those industries. Various carbonation processes had been developed, including direct carbonation and indirect carbonation with pH-swing, pressure-swing, and circular concept. Based on different purposes and sources of alkaline materials and CO_2 gas, different strategy of process development should be considered. Besides, obtained carbonated products and byproducts can give additional economic benefits. The utilization of products should be developed together with the establishment of the accounting methodology of CO_2 emissions reduction. Overall, mineral carbonation of industrial waste is an important route for industries to take as they move toward carbon neutrality and circular economy, as discussed in Chapter 4.

Photoelectrochemical reforming has been proven as a green and sustainable technology that can simultaneously mitigate waste and upcycle waste into valuable chemicals and clean fuels, with solar energy as the sole energy input, as indicated in Chapter 5. However, there remain fundamental challenges for the establishment of high-performance of the photoelectrochemical reforming devices, including light-harvesting capability, solubility and compatibility of the waste for the photoelectrochemical reforming, selectivity toward a specific product, and overall solar-to-chemical efficiency. With continued development and integration with other renewable technologies, photoelectrochemical reforming is expected to play an important role in enabling the circular and sustainable flow of materials and energy, and thus realizing a carbon-neutral future.

The nanophotonic platform based on precise optical design creates a novel fundamental study, such as exploration of light-induced dynamics and energy transfer, to practical applications like photovoltaic, photoelectrochemistry, and plasmonic photocatalysis. Addressing the issue of environmental sustainability in Chapter 6 is important for all human beings. The successful demonstration by the research community will guide us to prevail over the world in the goals of affordable and clean energy (SDG 7) and climate action (SDG 13).

Chapter 7 highlights the versatility and sustainability of water chestnut shell-derived carbon materials. These are inexpensive, ecofriendly electrode materials with significant potential applications in energy storage (as supercapacitors, sodium-ion batteries, and potassium-ion batteries) and power generation (such as microbial and plant microbial fuel cells). These materials address environmental challenges and promote cleaner and more efficient energy solutions, aligning with the goals of a circular economy.

Chapter 8 focuses on the potential of biomass energy. For biomass-to-energy conversions, at least three methods, including mechanical, thermal, and biological conversions, were developed. Among them, gasification and pyrolysis are the most popular because of their technological adaptability, high conversion effects, and potentially competitive costs. Mechanical methods were often used to pretreat biomass so that the following stages (i.e., the main method for the whole process of biomass conversion) can take place smoothly. Also, mechanical conversion alone can only be applied on a very small or individual scale because of its lack of competitiveness. In case of biological conversion, to optimize the efficiency of biofuel production, it is possible to combine two or more different methods. We update the latest research improvements of each method and delve into the analysis of each method to clarify the challenges and opportunities for organizations when they want to exploit biomass and produce biofuel on an industrial scale. Utilization of these energy sources not only ensures the energy security of countries, avoiding dependence on other countries rich in natural resources, but also resolves economic, social and environmental issues.

With the increasing demand for clean energy, hydrogen is considered as a potential alternative energy source. However, green hydrogen production currently accounts for less than 5% of total hydrogen production worldwide. This is an opportunity, as well as a challenge, for efforts to improve green hydrogen production processes to solve current energy problems. Currently, hydrogen production via thermal biomass conversion is considered as the most efficient and feasible for industrial scale production, as addressed in Chapter 9. Other approaches, such as biological biomass conversion or water splitting with low production efficiency, need to be improved to be competitive with fossil fuels as well as grey hydrogen.

Sustainable development has indeed become a crucial global issue. The need for environmentally friendly and sustainable energy solutions is paramount in our quest for a sustainable future. In this context, rSOC technology emerges as a promising option. By leveraging renewable sources like solar and wind power, rSOC systems can produce clean energy with a reduced carbon footprint, as investigated in Chapter 10. This enables us to maximize energy efficiency and reduce reliance on fossil fuels. As we strive to achieve sustainable development, rSOC technology presents an opportunity for greener and more sustainable energy solutions. Continued research, development, and investment in rSOC technology can further enhance its efficiency, cost-effectiveness, and applicability on a larger scale. By embracing rSOC technology and other sustainable energy solutions, we can play our part in mitigating climate change, protecting our environment, and securing a sustainable future for generations to come.

Chlorinated hydrocarbons are often used in industrial manufacturing. Due to their being heavier than water and having low solubility in water, they seep downwards when they infiltrate soil and groundwater due to gravity and form a pollution plume that spreads with the groundwater. Among various remediation methods, biological treatment is the most environmentally friendly. This is achieved through organohalide-respiring bacteria or the addition of commercial bacterial agents to enhance in-situ reductive dechlorination, as shown in Chapter 11. The effect of enhanced bioremediation is achieved by adding exogenous hydrogen. In order to comply with the global trend of carbon reduction, one direction is the use of biohydrogen. By immobilizing Dehalococcoides mccartyi BAV1 and Clostridium butyricum together in a silicone carrier, the reductive dechlorination effect of TCE and cis-1,2-DCE is enhanced. The biohydrogen produced by Clostridium butyricum is used for enhanced in-situ reductive dechlorination. Compared with in-situ hydrogen injection or increasing the H_2 in water by reacting nZVI with water, it is more in line with the United Nations Sustainable Development Goals, including SDG6 (Clean Water and Sanitation) and SDG13 (Climate Action). It is a potential choice for in-situ remediation.

Because the demand for batteries still increases, the number of waste batteries also soars. Thus, the recycling techniques of waste batteries are of utmost importance, as indicated in Chapter 12.

1. Environmental Protection: The toxic materials in batteries, such as lead, mercury, cadmium, and lithium, can contaminate soil, water, and air if not properly managed. Recycling ensures that these materials are safely extracted and disposed of, minimizing their environmental impact.
2. Resource Conservation: Batteries contain valuable resources like metals (e.g., lithium, cobalt, nickel) and plastics. Recycling allows for the recovery and reuse of these materials, reducing the need for raw material extraction and mining.
3. Energy Efficiency: Recycling batteries requires less energy compared to producing new batteries from virgin materials. The recycling process involves smelting or other techniques to extract valuable metals, which consumes less energy than the primary production of metals.

Overall, recycling waste batteries is becoming increasingly important, and advancements in recycling techniques are facilitating more efficient and sustainable management of battery waste. These trends reflect a shift towards a more circular and environmentally conscious approach to battery production and disposal.

In the future, recycling techniques for solar cell materials, as discussed in Chapter 13, are expected to see advancements in material separation technologies, such as automated robotic systems and advanced spectroscopic analysis, enabling more precise and efficient separation of different materials. Chemical recycling methods may gain prominence, utilizing solvents or chemical reactions to dissolve active layers and recover valuable materials like silicon, copper, silver, tin, indium, gallium, selenium, cesium, or lead. Advanced recovery techniques, including hydrometallurgical and pyrometallurgical processes, could be employed to extract and purify metals from solar cells. Designing solar modules with recyclability in mind, establishing specialized recycling facilities, or integrating solar cell recycling capabilities into existing electronic waste recycling plants may become more prevalent. Additionally, circular economy approaches can be adopted, promoting closed-loop systems where materials from decommissioned solar panels are recycled to manufacture new panels, reducing the need for virgin resources.

The importance of recycling materials from fuel cells lies in its multifaceted impact on environmental sustainability, resource conservation, and economic growth, addressed in Chapter 14. As the world faces the pressing challenges of climate change and resource depletion, recycling precious metals, rare earth elements, and other valuable components from fuel cells helps reduce the demand for virgin materials, curbing destructive mining practices and associated environmental degradation. This, in turn, lowers greenhouse gas emissions and lessens the burden on landfills, promoting a circular economy and minimizing waste. Furthermore, recycling fosters technological innovation and creates new economic opportunities in the emerging clean energy sector, driving job growth and economic development. Recent trends indicate a growing emphasis on research and development efforts to improve recycling methods, making the process more efficient and cost-effective. As fuel cell technology continues to advance and gain broader adoption in various industries, recycling materials is poised to play a pivotal role in promoting a sustainable future while meeting the increasing demand for clean and renewable energy solutions.

In this book, the journey through carbon capture, mineral carbonation, biomass energy, and the recycling of waste materials has illuminated the path toward a more harmonious coexistence between economic growth and ecological preservation. By fostering a culture of resource efficiency, waste reduction, and the integration of renewable energy, this book presents a blueprint for transformative change in our global approach to energy and resource management.

Looking ahead, the chapters within this book provide a glimpse into the potential of energy and resource management. The promising technologies discussed, point toward a future where our energy needs are met with minimal environmental impact. As we continue to refine and innovate these technologies, the prospect of achieving net-zero emissions becomes not just a goal but an attainable reality. Furthermore, the insights on waste recycling techniques for batteries, solar cells, and fuel cells signal a future where the life cycles of materials are extended, contributing to a circular economy. The book's overarching theme suggests a trajectory toward a more sustainable, interconnected, and environmentally conscious world. The future looks promising as these concepts evolve from theoretical frameworks to practical solutions, ultimately reshaping our relationship with the environment and setting the stage for a more harmonious coexistence between humanity and the planet.

Index

Pages in *italics* refer to figures and pages in **bold** refer to tables.

D

E

F

R

Y

Z